The History of Meteorology: to 1800

American Meteorological Society
Historical Monograph Series

The History of Meteorology: to 1800

The History of American Weather (All four volumes—$40.00)

Early American Hurricanes (1492-1870), 1963—$12.00

Early American Winters (Two volumes—$22.00)
(1604-1820), 1966—$12.00
(1821-1870), 1967—$12.00

Early American Tornadoes (1586-1870), 1970—$12.00

The Thermal Theory of Cyclones: A History of Meteorological Thought in the Nineteenth Century (In preparation)

The History of METEOROLOGY: to 1800

H. Howard Frisinger
Colorado State University

HISTORICAL MONOGRAPH SERIES

AMERICAN METEOROLOGICAL SOCIETY

Science History Publications
NEW YORK • 1977

Published jointly by

Science History Publications (USA)
a division of
Neale Watson Academic Publications, Inc.
156 Fifth Avenue, New York 10010

and the

American Meteorological Society
45 Beacon Street
Boston, Massachusetts 02108

(CIP data on final page)

Printed in the U.S.A.

Contents

To Mim

Preface

Within this century meteorology has emerged as one of the most active and important branches of the physical sciences. As man's activities have moved into the atmosphere and beyond, the science of the weather has assumed increasing significance in his life. Not only are giant steps being taken in the acquisition of meteorological knowledge, but extensive efforts are being devoted to finding ways of harnessing the tremendous energy present in the atmosphere. Such efforts as cloud seeding, hurricane modification, hail suppression, and fog dispersal exemplify these attempts to control the weather. As the importance of this science has increased, so has the interest in the history of meteorology. This interest has provided the motivation for this book.

Yet, notwithstanding the increased interest in the history of meteorology, very little has been written about it, and most of the work was done during the first quarter of this century by authors like Cleveland Abbe, Gustav Hellmann, and Sir Napier Shaw. The attempt to fill this gap even partially has occupied much of my interest and time over the past ten years.

There are, I believe, three major periods in the history of meteorology. The first, c. 600 B.C. to 1600 A.D., could be called "The Period of Speculation," when the dominant meteorological authority was Aristotle's *Meteorologica.*

The second period, 1600 A.D. to 1800 A.D., could be labeled "The Dawn of Scientific Meteorology." It was highlighted by the invention and early development of the basic meteorological instruments such as the thermometer, the barometer, and the hygrometer. Continuous and systematic meteorological observations, which were vital to the advancement of the science, were first initiated during this period. Thus, the foundations of modern meteorology were laid in the seventeenth and eighteenth centuries.

The third major period in the history of meteorology begins in 1800 and has been marked by the growth and refinement of its modern theories. It was after 1800 that meteorology, perhaps more than any other scientific discipline, drew extensively from related areas of the physical sciences for its growth and development. Such sister sciences as mathematics, physics, and chemistry have contributed much to advances in the study and understanding of the atmosphere. Because of the variance in the development of meteorology before and after 1800, I have chosen to restrict this study to its early history.

H. Howard Frisinger
Colorado State University

The Period of Speculation

Chapter One

The Beginning

As with any science, it is impossible to determine an exact beginning for meteorology. A distinction must be made between meteorology as a science and meteorology as a branch of knowledge. As we shall see, while meteorology as a science is comparatively young, as a branch of knowledge, it dates back to the origins of human civilizations.

As farmers and hunters, early men were strongly dependent on weather conditions, which forced them to watch atmospheric phenomena for signs that would foretell future weather. An accumulating collection of "weather signs" was handed down from generation to generation, gradually taking the form of short proverbs to facilitate their memorization.

Some of the earliest known writings contain fragmentary references to weather phenomena.[1] For example, *Job*, supposedly written near the beginning of the fifth century B.C.,[2] contains several speculations on the weather, including "Fair weather commeth out of the north."[3] Many of these weather proverbs are believed to have been ancient even at the time of their recording. Most are based on such weather phenomena as winds, clouds, and atmospheric optical conditions, although many others have a pronounced aspect of mystery or religion.

The first great civilizations of antiquity developed along some of the great rivers of Africa and Asia: the Nile in Africa, the Tigris and Euphrates in western Asia, the Indus and the Ganges in south-central Asia, and the Hwang Ho and the Yangtze in eastern Asia. Although new information about the ancient Chinese and Indian civilizations is being unearthed daily, most of our present knowledge of pre-Hellenistic civilizations is restricted to the Egyptians and the Babylonians.[4] In Egypt, meteorology had a marked religious character. As early as 3500 B.C. the Egyptians had a sky-religion with "rain-making" rituals.[5] As in all religions of antiquity, atmospheric phenomena were believed to be under the control of the gods.[6]

Babylonian civilization developed along the Tigris and

Euphrates valleys and flourished from approximately 3000 B.C. to 300 B.C.[7] Lacking suitable plants such as the papyrus used by the Egyptians, the Babylonians used clay as a writing surface (*Fig. 1.1*): With a stylus, inscriptions were pressed into clay tablets which were then baked to produce a permanent record.

Cuneiform Text
Fig. 1.1

The unearthing and deciphering of thousands of these tablets have shown the Babylonians as accomplished businessmen, mathematicians, and astronomers.[8] They have also indicated that meteorology developed a new character in Babylonian culture. For by trying to connect atmospheric phenomena with the movement of the heavenly bodies, the Babylonian astronomer-priests founded astrometeorology. Such forecasts as "When a dark halo surrounds the moon, the month will bring rain or will gather clouds," and "When a cloud grows dark in heaven, a wind will blow" resulted from their attempts to determine the relation between astronomy and meteorology.[9] The study of clouds, storms, winds, and thunder provided omens for good and bad events: "When it thunders in the day of the moon's disappearance, the crops will prosper and the market will be steady."[10]

The Babylonians had the wind rose of eight rumbs. They counted the four cardinal points in the order south, north, east, west (sutu, iltanu, sadu, amurra). Combining these words with (u), they formed the other points such as south-east (sutu u sadu), and north-west (iltanu u amurra). This method of combining the four principal winds to denote all others until recently had been thought to have originated much later, during the reign of Charlemagne.[11]

The first people to make regular meteorological observations and purposeful meteorological theories were the ancient Greeks. Early Greek cities were scattered throughout the eastern Mediterranean (*Fig. 1.2*), and it was in one of them, Miletus, in Asia Minor about 650 B.C., that the first known Ionian natural philosopher, mathematician, and meteorologist lived: Thales of Miletus, one of the "seven wise men" of antiquity.

Tradition lists Thales as the first natural philosopher to be credited with mathematical discoveries and proofs.[12] He was known to have been interested in meteorological phenomena: The early Greek historian Herodotus reports that around 585 B.C. Thales predicted a solar eclipse.[13] Following the lead of the Babylonians, whose works he apparently had studied, Thales attempted to associate weather phenomena with the movement of heavenly bodies. He is reported to have written on the equinox and the solstice, and to have studied the group of stars known as the Hyades, supposed by the ancients to indicate the approach of rain when they rose with the sun.[14]

Dating back to the beginning of recorded history, man has speculated on the basic elements which make up our world. Thales was no different. He contended that the world was compounded from one basic element—water,[15] which lay at the basis of all being, and by its mobility connected a life cycle which passed from the sky to the earth, through all living things, and then back to the sky again. Thus, Thales was aware of the cyclic movement of water falling from the sky in the form of rain, and condensing back into the sky. Although he was undoubtedly aware that clouds contained water, there has been no evidence to indicate that Thales understood the processes of condensation and cloud formation.

The Eastern Mediterranean in Classic Times

Fig. 1.2

An extensive traveler, Thales made at least one journey to Egypt, where he encountered a problem which had perplexed the Egyptians for centuries: Every year the Nile River would rise above its normal level and flood the surrounding areas. According to Seneca in his *Quaestiones Naturales,* Thales gave the following explanation:[16]

> The Etesian [northerly] winds hinder the descent of the Nile and check its course by driving the sea against its mouths. It is thus beaten back, and returns upon itself. Its rise is not the result of increase: it simply stops through being prevented from discharging, and presently, whenever it can, it bursts out into forbidden ground.

As will be seen, this problem continued to occupy the attention of natural philosophers for some three hundred years after Thales' time.[17]

Throughout the early history of meteorology the weather phenomena of thunder and lightning were a prominent topic for speculation by natural philosophers. Two of Thales' followers, Anaximander (ca. 611–547 B.C.) and Anaximenes (ca. 585–528 B.C.), had similar theories about the cause of thunder and lightning. They maintained that thunder was due to air smashing against clouds—that, as it rushed struggling through the cloud, it also kindled the flame of lightning.[18] This theory implied that there existed a fire-like substance in the atmosphere, a belief that prevailed in meteorological theories for well over 2,000 years.

Anaximander was a fellow citizen and companion of Thales. Late in his life Anaximander wrote a treatise, *peri physeós (de natura rerum),* the first on natural philosophy in the history of mankind.[19] Only a few lines of it, however, have come down to us. Anaximander was an acute observer of atmospheric phenomena, and it was probably this acuteness which led to his definition of winds as "a flowing of air."[20] He was the first to give this scientific definition of the wind, a definition that, surprisingly, was not generally accepted by subsequent natural philosophers for centuries.[21]

Anaximenes, another Miletan, accepted Thales' theory of a "basic" element underlying the world, but believed it to be air rather than water because of his observations upon the necessity of air for the support of life on earth.[22] He theorized that air contained an essence which he called "pneuma" and believed that this essence supported the universe in the same way as air supported human existence. Although air was the primary substance, it could take on all kinds of appearances by condensation or thickening, or by rarefaction or thinning. Changes in temperature were associated with these qualitative changes. Noting the curious result that air expelled through an open mouth is warm while air blown through nearly closed lips is cool, Anaximenes wrongly concluded that rarefaction increased the temperature while compression decreased it.[23]

The last of the celebrated philosophers of the Ionian School was Anaxagoras (ca. 499–427 B.C.), who taught at Athens, and became famous as a brilliant natural philosopher. The teachings and treatises of Anaxagoras, often called the "First Scientist," made him the standard ancient authority on natural philosophy.

Meteorology was among his numerous interests.[24] In fact, his scientific system was perhaps best exemplified in investigations of meteorological phenomena. His basic approach was first to observe nature carefully and then to devise experiments for testing hypotheses where simple observation had failed. One meteorological phenomenon which Anaxagoras sought to explain was summer hail, which had baffled those natural philosophers who thought that water could not freeze in warm weather.

Beginning with the observations that air temperature decreased with increasing height and that clouds contained moisture, Anaxagoras deduced that water would freeze at very high altitudes even in the summer. But how were the clouds to be forced up to these high altitudes? The question was easily answered by Anaxagoras who knew that heat caused objects to rise and created convection currents in the atmosphere. Thus, the heat of a summer day would drive the moisture-laden clouds to such an altitude that the moisture could freeze and fall back to the earth in the form of hail.[25]

To explain decreasing temperature with increasing altitude, Anaxagoras argued that increased height caused a progressive decrease in the intensity of sunlight reflected from the surface of the earth,[26] which in turn caused a decrease in air temperature. This temperature effect, however, only extended up so far. Beyond some point in upper space, the atmosphere began to assume a different form—that of a fire-like substance which Anaxagoras called "aether."[27] Hence, somewhere in the upper atmosphere the temperature would begin to rise, becoming burning hot. This reversal, however, occurred beyond altitudes which were meteorologically significant. Thus, for entirely false reasons, Anaxagoras had arrived at the true picture of the temperature-altitude relationship. Interestingly, this temperature-altitude relationship was not generally accepted in scientific circles until the nineteenth century.

Anaxagoras employed this theory of "aether" in the upper atmosphere to explain the cause of thunder and lightning. According to Aristotle, Anaxagoras theorized that there was fire in the clouds.[28] This fire was part of the aether which had descended into the lower atmosphere. The lightning was caused by this fire flashing through the clouds; thunder was the noise of the fire hissing when quenched by the moisture in the clouds. Why a hot substance in the atmosphere (the aether) could descend instead of ascend, as his theory on summer hail would indicate, was never answered by Anaxagoras.

Thales' and Anaxagoras' theories of a primary, universal element inspired numerous rivals.[29] The most accepted theory, which dominated meteorology for 2,000 years and to which even Aristotle adhered, was proposed by a Sicilian, Empedocles of Argigentum (ca. 492-430 B.C.).[30] Empedocles maintained that there were four basic elements in the universe: air, earth, fire, and water,[31] which were associated with four basic qualities: heat, cold, moisture, and dryness. Since water extinguishes fire, Empedocles concluded that these two elements were in opposition.

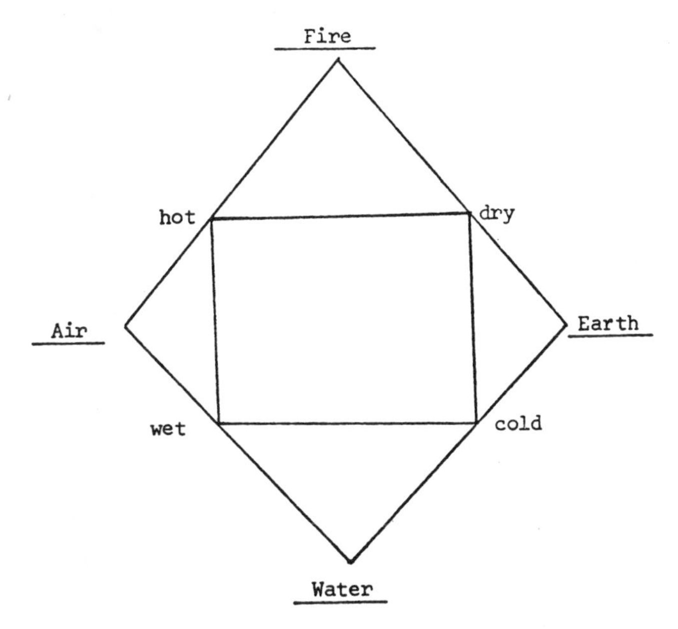

Empedocles' Concept of the Nature of the Universe
Fig. 1.3

Water and air, however, combined with each other or possessed an affinity for each other. Thus, the universe consisted of four basic elements associated with four primary qualities and the twin properties of opposition and affinity (*Fig. 1.3*).

Empedocles became interested in the causes of meteorological phenomena, an interest that included speculation on the cause of thunder and lightning. His theory was essentially the same as that of Anaxagoras except that Empedocles claimed that the fires in the clouds were rays of the sun that had become trapped in the clouds. He seems to have been the first to suggest that the lightning was engendered in the cloud.[32]

Applying his four-element concept of the universe, Empedocles attempted to employ the opposition of fire and water to explain the cause of the different climates—summer and winter. The basic elements fire and water continually opposed each other in the atmosphere. When the hot, dry fire gained the upper hand, summer resulted; when the wet, cold water gained mastery, winter resulted.[33] That these two basic elements moved about in a random manner did not explain why the seasons of summer and winter were so regular in their occurrence.

Another noted natural philosopher of the fifth century B.C. interested in weather phenomena was Democritus (ca. 460–370 B.C.), the atomist and geometer. He was according to his own boast, a great traveler: "I have wandered over a larger part of the earth than any other man of my time, inquiring about things most remote; I have observed very many climates and lands and have listened to very many learned men. . . ."[34] It was probably during a stay in Egypt that he considered the problem of the annual flooding of the Nile. Like Thales he pointed to the Etesian winds as the culprit. Democritus' explanation, however, was quite different from that of Thales and indicated a greater awareness of atmospheric activities. He maintained that the snow in the northern parts of the world melted at the time of the summer solstice and flowed away. Then clouds were formed by the vapors which, when driven towards the south, towards Egypt, by the Etesian winds, caused violent storms which filled the lakes and the Nile.[35]

While this indicates that Democritus accepted Anaximander's definition of winds as a flowing of air, it is more interesting as evidence that he might have partially sensed the important concept of the movement of storm systems, which was not generally appreciated until the eighteenth century. Previously it was thought that a particular storm did not move from one place to another. No evidence, however, has been found that Democritus actually reached this acute conclusion.

Democritus made use of his atomistic theory in his definition of wind[36] arguing that when there were many particles ("atoms") in a small empty space, wind was the outcome; on the other hand, when the space was large and the particles few, there was a "still, peaceful condition of the atmosphere." Seneca related the illustration used by Democritus:[37]

> To illustrate: in the market square or in a side street as long as there is a sprinkling of people there is no disturbance as one walks along it; but when a crowd meets in a narrow space, then they jostle against each other and quarreling arises. Similarly in this space which surrounds our earth; when many bodies have crowded a very small portion, it is unavoidable that they should jostle one another and be driven back and forward, and be intertwined and squeezed. Hence results wind: the particles that were struggling have had to give way, and after being tossed about and remaining in suspense for a long time they at length lean their weight toward one side. But when a few bodies occupy a large roomy place, they can neither ram each other nor be jostled by one another.

This theory was immediately criticized and discounted because it was often observed that wind did not always accompany a cloud-laden atmosphere.

The phenomenon of thunder and lightning also intrigued Democritus. On the basis of his atomistic theory, he explained thunder and lightning as the unequal mixture of particles, causing violent movements of, or within, clouds.[38] He believed that thunder and lightning occurred together, and were sensed separately because sight was quicker than hearing—a correct grasp of the simultaneity of thunder and lightning which was disregarded by subsequent natural philosophers and did not reappear for some 2,000 years.

Brief mention should also be made of the meteorological interest of Hippocrates of Cos (ca. 460–375 B.C.), often called the "Father of Medicine." An important part of his medical doctrine was the need to understand nature in order to understand the body and the soul of man. He believed that an understanding of meteorology was necessary to become a successful physician. In his treatise "On Airs, Waters, and Places," Hippocrates discussed different climates, the effect of these climates on the health of the inhabitants, and the diseases prevalent in localities characterized by their exposure to particular winds.[39] Hippocrates, however, apparently did not make any attempts to speculate on the causes of the meteorological phenomena he studied. His studies were focused only on the effect of weather on the health of man.

The last natural philosopher in the pre-Aristotelian period known to have taken an interest in meteorology was Eudoxus of Cnidos (ca. 408-355B.C.), a pupil of Plato. Eudoxus is believed to be the author of a treatise (*Ceimonos Prognostica*) on bad-weather predictions,[40] which was of Babylonian origin.[41] Also he had an interesting theory on the periodicity in weather phenomena. Pliny, in his *Natural History*, stated that Eudoxus claimed that there was a regular recurrence of all meteorological phenomena.[42] This periodicity occurred not only in winds, but in other sorts of bad weather as well. This recurrence was in four-year periods, and a period always began in a leap year at the rising of the star Sirius.[43]

In this early stage of meteorological development, many natural philosophers devoted their attention to the study of weather. Their studies, however, had to be made without the benefit of meteorological instruments, which were not developed until the seventeenth century A.D. Consequently, the study of meteorological phenomena by the scientists of antiquity was largely qualitative rather than quantitative. The theories proposed, with the possible exceptions of cloud movements, were not verified by measurements of the phenomena they concerned. They could be accepted or rejected only by being related speculatively to more general theories of nature such as the theory of the four elements. Nevertheless, these theories formed the basis for Aristotle's *Meteorologica* which in turn became the unquestioned authority on weather theory for the next 2,000 years.[44]

References

1. Harvey A. Zinszer, "Meteorological Mileposts," *Scientific Monthly* 58 (1944): 261.

2. *The Jerusalem Bible* (New York: Doubleday & Company, Inc., 1966), p. 727.

3. *Job*, Ct. 37, verse 22.

4. William Dampier, *A History of Science* (New York: The Macmillan Company, 1935), p. 1.

5. G.A. Wainwright, *The Sky Religion in Egypt* (Cambridge: The University Press, 1938), p. 11.

6. According to the Greeks, Zeus governed the meteors; the symbol of his power being the lightning. See L. Dufour, "Les grandes époques de l'histoire de la météorologie," *Ciel et Terre* 59 (1943): 356.

7. For a thorough discussion of the Babylonian civilization with emphasis on the science of that period, see O. Neugebauer, *The Exact Sciences in Antiquity* (New York: Harper Brothers, 1962).

8. George Sarton, *A History of Science: Ancient Science Through the Golden Age of Greece* (Cambridge: Harvard University Press, 1960), pp. 74-94.
9. Gustav Hellmann, "The Dawn of Meteorology," *Quarterly Journal of the Royal Meteorological Society* 34 (1908): 223.

10. *Ibid.*

11. *Ibid.*, p. 224.

12. Howard Eves, *An Introduction to the History of Mathematics* (New York: Holt, Rinehart and Winston, 1969), p. 50.

13. David Eugene Smith, *History of Mathematics* (New York: Dover Publications, Inc., 1958), p. 67.

14. Kathleen Freeman, *The Pre-Socratic Philosophers*, 3rd ed. (Oxford: Basel Blackwell, 1953), p. 51.

15. W. S. Fowler, *The Development of Scientific Method* (London: Pergamon Press, 1962), p. 5.

16. Seneca, *Quaestiones Naturales*, trans. John Clarke (London: Macmillan and Co., Ltd., 1910), p. 174.

17. See H. Howard Frisinger, "Early Theories on the Nile Floods," *Weather* 20 (1965): 206-208.

18. Seneca, *op. cit.*, p. 67.

19. Sarton, *op. cit.*, p. 173.

20. Sigmund Gunther and W. Windelband, *Geschichte Der Antiken Naturwissenschaft und Philosophie* (Nordlingen: Verlag Der C. H. Beckschen Buchhandlung, 1888), p. 143.

21. Hellmann, *op. cit.*, p. 224.

22. Fowler, *op. cit.*, p. 5.

23. The opposite, in fact, is true. Adiabatic compression increases the temperature, while adiabatic dilatation decreases it.

24. Daniel E. Gershenson and Daniel A. Greenberg, *Anaxagoras and the Birth of Scientific Method* (New York: Blaisdell Publishing Co., 1964), p. 3.

25. Aristotle, *Meteorologica*, trans. H.D.P. Lee (Cambridge, Mass.: Harvard University Press, 1952), p. 81.

26. Gershenson and Greenberg, *op. cit.*, p. 44.

27. *Ibid.*

28. Aristotle, *op. cit.*, p. 227.

29. For a detailed discussion of these different theories, see René Taton (ed.), *Histoire Génèrale Des Sciences* (Paris: Presses Universitaires De France, 1957), 1: 210–219.

30. Sarton, *op. cit.*, p. 247.

31. Fowler, *op. cit.*, p. 7.

32. Aristotle, *op. cit.*, p. 227.

33. Kathleen Freeman, *Ancilla to the Pre-Socratic Philosophers* (Oxford: Basil Blackwell, 1948), pp. 51–68.

34. David E. Smith, *History of Mathematics* (New York: Dover Publications, Inc., I, 1958), p. 81.

35. M.R. Cohen and I.E. Drabkin, *A Source Book in Greek Science* (New York: McGraw-Hill Co., Inc., 1948), p. 383.

36. Seneca, *op. cit.*, p. 195.

37. *Ibid.*

38. Freeman, *op. cit.*, pp. 304–306.

39. Hippocrates, *The Genuine Works of Hippocrates*, trans. Francis Adams (New York: William Wood and Co., I, 1886), pp. 156–183.

40. Text in *Catalogus Codicum Astrologorum Graecorum* 7 (1908): 183–187.

41. Sarton, *op. cit.*, p. 447.

42. Secundus Plinius, *Pliny: Natural History*, trans. H. Rackham (London: William Heinemann Ltd., 1938), 1: 269.

43. For an account of some of the more modern theories on weather cycles, from weekly to fifty-year cycles, see Richard Gregory, "Weather recurrences and weather cycles," *Quart. Jour. of the Roy. Meteor. Soc.* 41 (1930): 103–120.

44. Zinszer, *op. cit.*, p. 261.

Thales of Miletus
(c. 650 B.C.)

Aristotle (384–322 B.C.)

Chapter Two

Aristotle's *Meteorologica*

Aristotle (384–322 B.C.) was born at Stagira, a Greek colony a few miles from the present monastic settlement of Mount Athos. He moved to Athens and at seventeen became a pupil of Plato. Following his master's death in 347 B.C., he settled in Lesbos, an island off the coast of Asia Minor. His rising reputation as a scholar resulted in an appointment as tutor to the young prince Alexander of Macedon, the future Alexander the Great. Returning to Athens in 336 B.C., Aristotle became one of the most noted public teachers, writers, and philosophers of all time. Included among his many works is the treatise *Meteorologica,* the inspiration for the growth of meteorology into a science.[1]

Written around 340 B.C., Aristotle's *Meteorologica* is the oldest comprehensive treatise on the subject of meteorology.[2] The work is in four books, of which the first three deal with what we now consider meteorology, while the fourth book is mainly about chemistry.[3] Subjects considered in the first three books include: the formation of rain, clouds and mist, hail, winds, climatic changes, thunder and lightning, and hurricanes.

The arguments in *Meteorologica* were founded on two basic theories. First, Aristotle believed that the universe was spherical in form. He accepted the system of Eudoxus which accounted for the movements of the stars and planets by a system of concentric spheres whose combined motions produced the apparent movements of the heavenly bodies.[4] The earth was the inner core of these concentric spheres which were formed by the orbits of the heavenly bodies. Aristotle divided the universe into two major regions: the celestial region beyond the orbit of the moon, and the terrestrial or sublunar region-sphere of the moon's orbit about the earth. Thus, he made a precise distinction between astronomy and his new subject, meteorology. The former was restricted to the celestial region (including the orbit of the moon); the latter was restricted to phenomena of the terrestrial region.

His second basic theory was the "four-element theory" of Empedocles.[5] Aristotle pictured the terrestrial region as consisting of four elements—earth, water, air, and fire—arranged in concentric spherical strata with the earth at the center (*Fig. 2.1*). This stratification, however, was not rigid. Dry land rose above water, and fire often burned on the earth. Also, all elements were thought to be in constant processes of interchange, one into the other. When the heat of the sun reached the earth's surface, it mixed with the cold and moist water to form a new substance, warm and moist, essentially like air. The sun's heat similarly acted upon the cold and dry earth to produce another substance, warm and dry, essentially like fire.

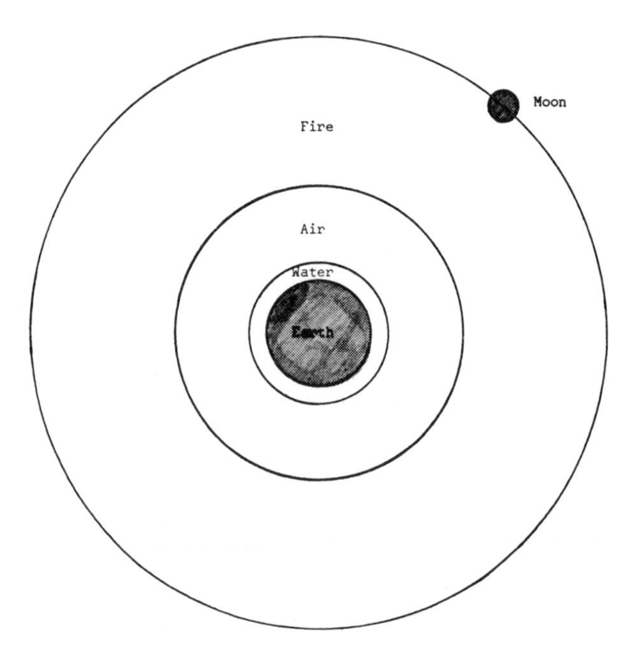

Aristotle's Concept of the Universe
Fig. 2.1

Thus, the sun drew up two kinds of "evaporations." One was the wet and generally warm vapors, which resulted in such phenomena as clouds, rain, etc. The other was the hot and dry vapors which provided source material for such phenomena as wind, thunder, etc.

There were thus two main strata in Aristotle's atmosphere, air and fire. But within the sphere of air there were certain further differentiations. Aristotle theorized that clouds could form neither beyond the tops of the highest mountain, because the air above the mountains contained "fire" and was carried round with the celestial motion, nor close to the earth, because the heat reflected from the earth also prevented cloud formation. There was, therefore, a strata in the atmosphere between the height of the highest mountains and the earth's surface in which clouds could form (*Fig. 2.2*).[6]

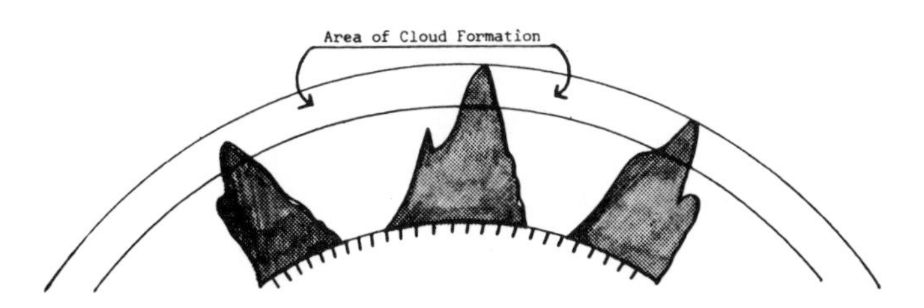

Aristotle's Theory of Cloud Formation
Fig. 2.2

Meteorologica also contains an accumulation of facts collected from former natural philosophers, historians, poets, and common experience.[7] A number of weather prognostications in this work are derived from the Egyptians and much of the work is of definite Babylonian origin, especially the classification of the winds.[8] Hence, *Meteorologica* represented the sum of the meteorological knowledge of its time. An excellent example of Aristotle's development of a meteorological theory from well-known facts is his discussion of hail:

> In considering the process by which hail is produced, we must take into account both facts whose interpretation is straightforward and those which appear to be inexplicable.
>
> Hail is ice, and water freezes in the winter: yet hailstorms are commonest in spring and autumn, rather less common at the end of the summer, and rare in winter when they occur when it is not very cold. And, in general, hailstorms occur in milder districts, snowstorms in colder.
>
> It is also odd that water should freeze in the upper region; for it cannot freeze before it becomes water, and yet having become water it cannot remain suspended in the air for any length of time. Nor can we maintain that just as drops of water ride aloft because of their minuteness and rest on the air, like minute particles of earth or gold that often float on water, so here the water floats on the air till a number of the small drops coalesce to form the large drops that fall.

This cannot take place in the case of hail, because frozen drops cannot coalesce like liquid ones. Clearly then drops of water of the requisite size must have been suspended in the air: otherwise their size when frozen could not have been so large.

Some of them think that the cause of the origin is as follows: when a cloud is forced up into the upper region where the temperature is lower because reflection of the sun's rays from the earth does not reach it, the water when it gets there is frozen: and so hailstorms occur more often in summer and in warm districts because the heat forces the clouds up farther from the earth.[9] But in the very high places hail falls very infrequently; but on their theory this should not be so, for we can see that snow falls mostly in high places. Clouds have often been seen swept along with a great noise close to the earth, and have struck fear into those that heard and saw them as portents of some greater catastrophe. But sometimes, when such clouds have been seen without any accompanying noise, hail falls in great quantities and the stones are of an incredible size, and irregular in shape; the reason being that they have not had long to fall because they were frozen close to the earth, and not, as the theory we are criticizing maintains, far above it. Moreover, large hailstones must be formed by an intense cause of freezing; for it is obvious to everyone that hail is ice. But hailstones that are not rounded in shape are large in size, which is a proof that they have frozen close to the earth: for stones which fall farther are worn down the course of their fall and so become round in shape and smaller in size.

It is clear then that the freezing does not take place because the cloud is forced up into the cold upper region.

Now we know that hot and cold have a mutual reaction on one another (which is the reason why subterranean places are cold in hot weather and warm in frosty weather). This reaction we must suppose takes place in the upper region, so that in warmer seasons the cold is concentrated within the surrounding heat. This sometimes causes a rapid formation of water from cloud. And for this reason you get larger raindrops on warm days than in winter and more violent rainfall—rainfall is said to be more violent when it is heavier, and a heavier rainfall is caused by rapidity of condensation. (The process is just the opposite of what Anaxagoras says it is: He says it takes place when clouds rise into the cold air: we say it takes place when clouds descend into the warm air and is most violent when the cloud descends farthest.) Sometimes, on the other hand, the cold is even more concentrated within by the heat outside it, and freezes the water which it has produced, so forming hail. This happens when the water freezes before it has time to fall. For if it takes a given time to fall, but the cold being freezes it in a lesser time,

> there is nothing to prevent it freezing in the air, if the time taken to freeze it is shorter than the time of its fall. The nearer the earth and the more intense the freezing, the more violent the rainfall and the larger the drops or the hailstones because of the shortness of their fall. For the same reason large raindrops do not fall thickly. Hail is rarer in the summer than in spring or autumn, though commoner than in winter, because in summer the air is drier: but in spring it is still moist, in autumn it is beginning to become so. For the same reason hailstones do sometimes occur in late summer, as we have said. If the water has been previously heated, this contributes to the rapidity with which it freezes: for it cools more quickly. (Thus so many people when they want to cool water quickly first stand it in the sun: and the inhabitants of Pontus when they encamp on the ice to fish—they catch fish through a hole which they make in the ice—pour hot water on their rods because it freezes quicker, using the ice like solder to fix their rods.) And water that condenses in the air in warm districts and seasons gets hot quickly.
>
> For the same reason in Arabia and Aethiopia rain falls in the summer and not in the winter, and falls with violence and many times on the same day: for the clouds are cooled quickly by the reaction due to the great heat of the country.
>
> So much then for our account of the causes and nature of rain, dew, snow, hoar frost and hail. (Book I, Ch. 12)

This discussion illustrates the method used by Aristotle throughout his treatise. He was very fond of introducing his theories by first presenting those of others, and then refuting them. The differing opinions between Anaxagoras and Aristotle on the formation of hail illustrate a basic difference between Aristotle's method of developing a meteorological theory, and that of his predecessors.

Anaxagoras and the other earlier natural philosophers were largely inductive in their approach to speculations on weather phenomena: Their theories were based largely on their observations. Aristotle, however, employed a more deductive approach, explaining various weather phenomena on the basis of his preconceived meteorological theories. Instead of using weather observations to develop his theories, Aristotle frequently interpreted these observations in such a way as to support preconceived beliefs. This was often accomplished by employing arguments by analogies where the analogies were assumed rather than demonstrated (for example, that concerning the temperature in subterranean places, in the discussion of hail). The extent to which Aristotle applied deductive reasoning was very evident in his discussion of winds.

As usual, he began by refuting the opinion of Anaximander and others that wind was simply a moving current of air. Recalling his theory that the sun drew up two types of exhalations from the earth, Aristotle claimed that the origin of wind was the dry, hot exhalation. He explained the cause of wind with an analogy of rivers which represent the gradually accumulated flow of water from the mountains downwards. In the same way wind was due to the gradual accumulation of the dry, hot exhalation from the earth. Thus:

> The facts also make it clear that winds are formed by the gradual collection of small quantities of exhalation, in the same way that rivers form when the earth is wet. For they are all least strong at their place of origin, but blow strongly as they travel farther from it. Besides, the north, that is the region immediately about the pole, is calm and windless in winter; but the wind which blows so gently there that it passes unnoticed, becomes strong as it moves farther afield. (Book I, Ch. 4)

Aristotle explained that the winds blow horizontally, although the exhalation rises vertically, "because the whole body of air surrounding the earth follows the motion of the heavens."[10]

According to Aristotle, there were two main winds: from the north and from the south. The north winds emanated from the cold regions under the Great Bear, the northern limit of the habitable world, and thus were cold. Those from the south came, not from the South Pole, but from the tropic of Cancer—the southern limit of the habitable world since beyond it the heat was believed to be too great for life. Because of the region from which they emanated, the south winds were hot winds.

The Aristotelian wind classification was based on his meteorological theory of their connection with the sun. The Greeks of that time had very limited means of expressing directions. As a consequence, Aristotle employed such astronomical directions as equinoctial sunrise, winter sunset, midday sun, etc., to indicate the directions of the various winds. Dividing the compass-card into twelve equal sectors, he enumerated the different winds and their directions (*Fig. 2.3*). This duodecimal division strongly suggests a Babylonian origin. Aristotle noted there was no opposite wind to the wind Meses (K), nor to the wind Thrascias

(I), "except perhaps a local wind called by the inhabitants Phoenicias." He explained the greater number of northerly winds as follows:

> First, our inhabited region lies towards the north; second, far more rain and snow is pushed up into this region because the other lies beneath the sun and its course. These melt and are absorbed by the earth and when subsequently heated by the sun and the earth's own heat cause a greater and more extensive exhalation. (Book II, Ch. 6)

The ever popular phenomena for speculation, thunder and lightning, were also attributed by Aristotle to the dry exhalation in the atmosphere. He claimed that the dry exhalation became

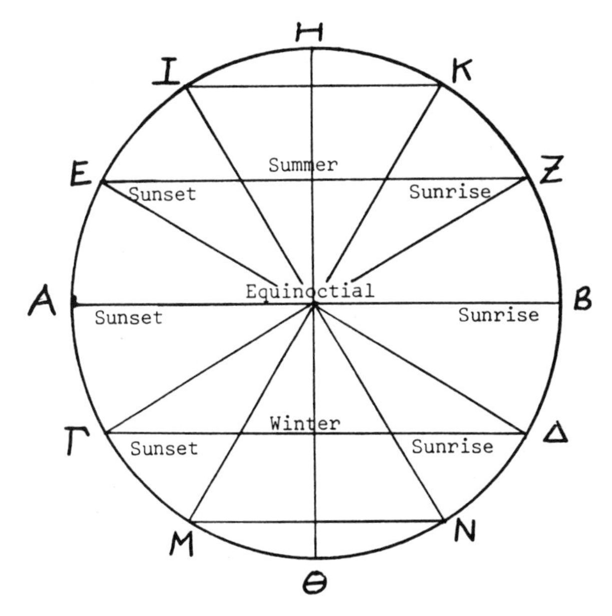

Aristotle's Wind Rose

Fig. 2.3

trapped in clouds and was forcibly ejected as the clouds condensed, striking the surrounding clouds and causing the thunder. Different sounds were produced because of the lack of uniformity in cloud composition. This ejected "wind" then burned with a "fine and gentle fire, and it is then what we call lightning."[11] Contrary to earlier opinion, Aristotle claimed that lightning followed thunder.

As has been seen, Aristotle based his explanations of weather phenomena on the theory that there were two kinds of exhalation, moist and dry, and that their combination (air) contained both potentially. Clouds also contained both exhalations. He applied this theory to explain yet another weather phenomenon, hurricanes, in an interesting way:

> The windy exhalation causes thunder and lightning when it is produced in small quantities, widely dispersed, and at frequent intervals, and when it spreads quickly and is of extreme rarity. But when it is produced in a compact mass and is denser, the result is a hurricane, which owes its violence to the force which the speed of its separation gives it. (Book III, Ch. 1)

Here Aristotle was explaining the violent winds associated with a hurricane, not its accompanying rain which was due to the moist exhalation.

Aristotle's treatise provides an excellent example of the many errors made by the Greeks in natural science because of their failure to develop an experimental scientific method; a major reason for this failure was the absence of precision instruments. Developed theory or natural philosophy tended to outweigh experimental evidence. *Meteorologica* was the product of the natural philosopher, not the natural scientist. Nevertheless, it is of great importance in the history of meteorology. It is the earliest known effort at a systematic discussion of meteorology, and was the unquestioned authority on weather theory for over 2,000 years.[12] In fact, all textbooks on meteorology in Western civilization up to the end of the seventeenth century were based exclusively on Aristotle's *Meteorologica.*[13]

References

1. For a discussion of the pre-Aristotelian development of meteorology, see H. Howard Frisinger, "Meteorology Before Aristotle," *Bulletin of the American Meteorological Society* 52, No. 11 (November, 1971): 1078–1080.

2. Carl B. Boyer, *The Rainbow* (New York: Thomas Yoseloff, 1959), p. 38.

3. There are now several English translations of this work. The one used here was from the translation by H.D.P. Lee; see Aristotle, *Meteorologica*, trans. H.D.P. Lee (Cambridge: Harvard University Press, 1952).

4. Aristotle, *De Caelo*, trans. W.K.C. Guthrie (Cambridge: Harvard University Press, 1939), p. 213.

5. W.S. Fowler, *The Development of Scientific Method* (London: Pergamon Press, 1962), p. 7.

6. Aristotle, *Meteorologica*, trans. H.D.P. Lee (Cambridge: Harvard University Press, 1952) pp. 6-9, 27.

7. William Napier Shaw, *Manual of Meteorology* (Cambridge: The University Press, 1926), 1: 76.

8. Alexander Buchan, *Handy Book of Meteorology* (London: William Blackwood and Sons, 1868), p. 2; D'Arcy Thompson, "The Greek Winds," *Classical Review* 32 (1918): 53.

9. This is the theory of Anaxagoras, as Aristotle later informs us.

10. Note that in Hellenic times the earth was considered the center of the Universe, and the other heavenly bodies moved around the earth.

11. Aristotle, *Meteorologica*, p. 225.

12. Harvey A. Zinszer, "Meteorological Mileposts," *Scientific Monthly* 58 (1944): 261-264.

13. Gustav Hellmann, "The Dawn of Meteorology," *Quarterly Journal of the Royal Meteorological Society* 34 (1908): 228.

Roger Bacon (1214-1294)

Chapter Three

The Dark Before the Dawn

For 2,000 years after Aristotle, there was very little progress in meteorology. Considerable efforts were devoted to atmospheric optics, but most of the scant attention given to meteorology consisted of commentaries on Aristotle's treatise. Most of his distinguished successors added little to the perfection of his system. There were a few, however, who did expand on his theories, especially in those areas to which he had paid slight attention. One was his pupil, Theophrastus of Eresos, who wrote "De Signis Tempestatum" (*On Weather Signs*) and a treatise, "De Ventis," on winds. The practice of weather prognostication by empirical rules dates from these treatises. Theophrastus gives some eighty different signs of rain, forty-five of wind, fifty of storm, twenty-four of fair weather, and seven signs of weather for periods of a year or less. In looking at the overall weather picture, Theophrastus advocates a general principle:[1]

> Now the first point to be seized is that the various periods are all divided in half, so that one's study of the year the month or the day should take account of those divisions. The year is divided in half by the setting and rising of the Pleiad; for from the setting to the rising is a half year. So that to begin with the whole period is divided into halves: and a like division is effected by the solstices and equinoxes. From which it follows that, whatever is the condition of the atmosphere when the Pleiad sets, that it continues in general to be till the winter solstice, and, if it does change, the change only takes place after the solstice: while, if it does not change it continues the same till the spring equinox: the same principle holds good from that time to the rising of the Pleiad, from that again to the summer solstice, from that to the setting of the Pleiad.
>
> So too is it with each month; the full moon and the eighth and the fourth days make divisions into halves, so that one should make the new moon the starting point of one's survey. A change most often takes place on the fourth day, or, failing that, on the eighth, or, failing that, at the full moon; after that the periods are from the full moon to the eighth day from the end of the month, from that to the fourth day from the end, and from that to the new moon.

> The divisions of the day follow in general, the same principle: there is the sunrise, the mid-morning, noon, mid-afternoon, and sunset; and the corresponding divisions of the night have like effects in the matter of wind storms and fair weather; that is to say, if there is to be a change, it will generally occur at one of these divisions.

Applying the above principle and that of a general balance in the yearly weather, Theophrastus comes up with such general forecasts as "If a great deal of rain falls in winter, the spring is usually dry; if the winter has been dry, the spring is usually wet" and "If the autumn is unusually fine, the succeeding spring is general cold."[2]

Unusual actions of animals and birds have for centuries been observed as indications of future weather. The still popular empirical rules, based on the actions of animals and birds, are enumerated in this treatise. Such rules as "It is a sign of storm or rain when the ox licks his fore-hoof," "A dog rolling on the ground is a sign of violent storm," and "It indicates an early winter when the breeding season of sheep begins early" are quite familiar to many a present-day farmer.

The rules for forecasting from observation of certain astronomical and atmospheric phenomena also appear in Theophrastus' work. Many shooting stars are considered a sign of rain or wind. Observing the moon is also important, i.e., "If the moon looks fiery, it indicates breezy weather for that month, if dusky, wet weather." It is interesting to observe that nearly all of the still popular empirical rules for weather forecasts ("Reddish sky at sunrise foretells rain") come from this short work.

Theophrastus made no attempt to explain the different atmospheric phenomena, but referred all such consideration to the Aristotelian method. While Aristotle's work was, thus, largely theoretical, Theophrastus' short treatise was completely practical. It is the oldest collection of weather signs to have survived, and most later collections were based on it.

Speculation on the cause of the annual Nile flooding was temporarily put to rest by the famous mathematician and geographer Eratosthenes (ca. 274–194 B.C.), who, according to Proclus, declared that it was definitely known that man had gone to the sources of the Nile and observed the rains.[3] Although the real cause of this annual phenomenon was now determined, interest in its meteorological implications has continued up to the twentieth century.[4]

The Golden Age of Greek science ended during the first century B.C. with Roman dominance of the eastern Mediterranean. The Romans cared not at all for pure science, which to them always remained an exotic endeavor: Science was only important if it had practical applications. Consequently, mathematics and the physical and natural sciences declined radically. One of the few noted natural scientists of this period was Posidonius (135-50 B.C.), who pursued physical investigations with considerable zeal. He was interested in meteorological speculations and closely followed the theories of Aristotle. Thus, Posidonius theorized that thunder was the bursting of the dry exhalation that had become trapped in the clouds. Although most ancient natural philosophers held that the maximum cloud and wind height extended up to 111 miles, Posidonius claimed that winds and clouds only reached up to a height of around five miles, beyond which the air was clear and liquid and perfectly luminous.[5] Aristotle's influence was clearly evident.

By the end of the second century B.C., the center of scientific activity was not Athens, but the city of Alexandria, which Alexander the Great had founded in the Nile Delta. The greatest of the ancient world libraries and the first international university were established there.

One of the many scientists whose name is connected with Alexandria was Claudius Ptolemy, or Ptolemy the Astronomer (ca. 85-165 A.D.). His *Almagest* is antiquity's most sophisticated astronomical treatise. Since the ancients generally considered meteorology a branch of astronomy, it is not surprising that Ptolemy displayed interest in weather phenomena and forecasting. In the *Tetrabiblos,* Ptolemy gave several astrological weather prognostications like the following based on the appearance of the moon:[6]

> We must observe the moon in its course three days before or three days after new moon, full moon, and the quarters. For when it appears thin and clear and has nothing around it, it signifies clear weather. If it is thin and red, and the whole disk of the unlighted portion is visible and somewhat disturbed, it indicates winds, in that direction in which it is particularly inclined. If it is observed to be dark, or pale, and thick, it signifies storms and rains.

For the next 1,000 years, this treatise of Ptolemy's was the basic authority for astrological weather prediction.[7]

Ptolemy was also a noted geographer, and constructed a map of the world (*Fig. 3.1*) divided into climatic zones which were classified solely with reference to their conditions of illumination in which the length of the longest day increased successively by half an hour between the equator and the Arctic Circle. These zones were therefore of different widths. The first climate, on the equator, embraced eight and one-half degrees of latitude, while the twenty-fourth climatic zone, at the Arctic Circle, embraced only three minutes of latitude. Ptolemy discussed the general temperature variations between these zones.[8]

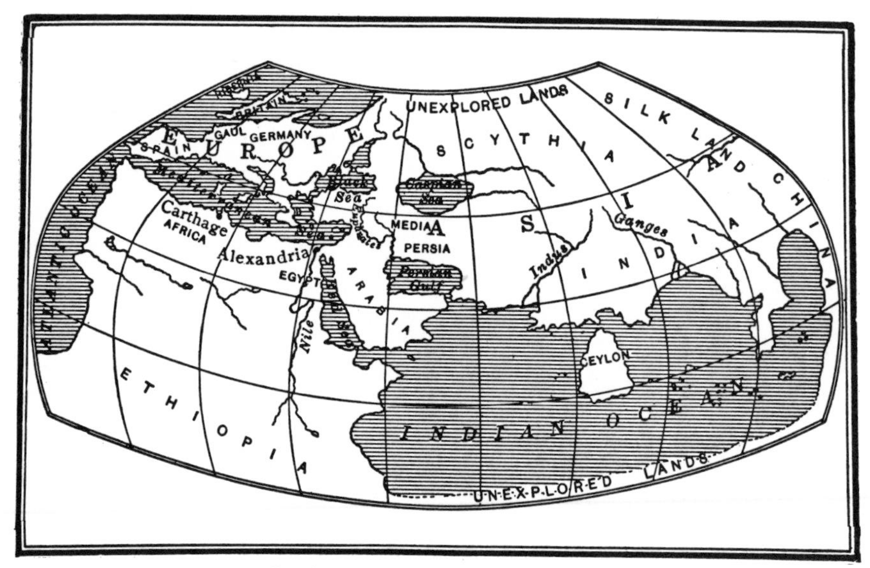

Ptolemy's Map of the World

Fig. 3.1

Ptolemy's division of the earth into climatic zones was made purely on an astronomical basis. As temperature is a principal element in any climatic classification, and as temperature depends to a large extent on the incoming solar radiation, there was some justification for such a division. Other important elements, however, such as precipitation, may vary considerably within such a climatic zone. Hence, the astronomical method of classifying climates has been refined in recent years, although remnants of the old division still appear, e.g., the Equatorial Zone and the Polar Zone.

Although the early Greek natural philosophers adopted much from earlier civilizations, e.g., the Babylonian eight-part wind rose used in the octagonal Tower of the Winds in Athens

(*Fig. 3.2*), a great debt is owed them.[9] They began the scientific inquiry into the different weather phenomena and their various interrelations. Moreover, these early scientists developed the most compendious expositions of the practical application of weather knowledge in the history of meteorology up to modern times.[10]

Tower of the Winds in Athens
Fig. 3.2

We should not leave this period without a brief mention of the Roman commentators whose works preserved many of the Greek theories in the natural sciences. For the history of meteorology, the two most important commentators were Lucius Seneca (3 B.C.-65 A.D.) and Secundus Pliny (Pliny the elder) (23-79 A.D.).

Near the end of his life (63-64) Seneca wrote *Quaestiones Naturales*. Dealing chiefly with astronomy and meteorology, the treatise combined the findings of Roman science with those of the Greeks, Babylonians, and Egyptians. Seneca covered the whole realm of meteorological phenomena from winds to thunder and lightning. He always sought to give fair treatment to different proposed meteorological theories, and his own conclusions were often compromises between them. Thus, for example, Seneca considered wind the result not only of the atmosphere in motion (Anaximander), but also of "evaporation" from land (Aristotle).

Seneca was a careful observer of the weather, and there is some evidence in his work that he possessed a degree of true scientific spirit and imagination. Seneca, however, was a moralist first and a scientist afterwards. Science led to theology and thus had a direct bearing on man's destiny and fate. As the atmosphere joins heaven and earth, atmospheric phenomena such as lightning were tied to fate. In the early development of meteorology the importance of Seneca's *Quaestiones Naturales* was its compilation of ancient Greek meteorological theories.[11]

Pliny's most important scientific work was his *Natural History*[12] which was compiled from some 2,000 works by 146 Roman and 326 Greek authors, most of which are now lost. The second book, or chapter, is on meteorology. Pliny noted that since the earliest times there had been persistent meteorological study and that more than twenty Greek authors had published observations about meteorological subjects. He discussed the various theories of former natural philosophers but failed to make any contributions himself. Thus, like Seneca's *Quaestiones Naturales,* Pliny's *Natural History* is significant for preserving earlier speculation on meteorological subjects.

The fall of the Roman Empire in the West was the beginning of a long, quiet period in the development of meteorology. From 400–1100, however, study never completely ceased. Some scientific spirit, as well as Greek and Latin learning, was preserved by the clergy. One of the greatest of these medieval church scholars was Bede the Venerable (ca. 673–735), the first Englishman to write on the weather, called by Botley "a founder of English meteorology."[13] Among his scientific works was *De Natura Rerum* written around 703. In this treatise are chapters devoted to the atmosphere, wind, thunder, lightning, clouds, and snow.[14] These meteorological chapters consisted of a summary of the knowledge then available obtained chiefly from classical sources. There is a familiar classical "ring" to such theories as wind being "air disturbed or in motion as may be proved with a fan" and "thunder is generated by the clashings of clouds driven by the winds which are conceived among them." Bede made no reference to the theories of Aristotle since his works were unknown in Western Europe before the twelfth century. That Bede's meteorological discussions were not always free of superstition is evidenced by his claim that thunder with a west wind signified "a very bad pestilence."[15] On the whole, however, since Bede's *De Natura Rerum* was an attempt to treat meteorology in a less philosophical, more scientific light, it did signal some advance.

Bede was not the only medieval scholar to display an interest in meteorology. At the beginning of the seventh century, there was no abler prelate in Christendom than Isidore, Bishop of Seville (ca. 570–636).[16] In such works as *Etymologiae, De Ordine Creaturarum,* and *De Natura Rerum,* this noted Spanish scholar devoted a considerable amount of attention to meteorological questions.[17] Like Bede, Isidore's thinking was hampered by the then prevailing theological views of science. He displayed, however, considerable power of thought; indeed, when he discussed such weather phenomena as frost, rain, hail, and snow, his theories were rational and gave evidence that if he could have broken away from his adhesion to the letter of Scripture, he might have given a great impetus to the evolution of meteorology.

The Moslems' major role in the history of science was as the "caretaker" of ancient cultures. They translated many Hindu and Greek works, including Aristotle's *Meteorologica,* into Arabic, which were later to be retranslated into Latin by Western scholars.[18]

The greatest Moslem physicist and one of the most noted students of optics of all time was Ibn Al-Haitham (Alhazen) (ca. 965–1039).[19] The Latin translation of his main work, *Opticae Thesaurus,* exerted a great influence upon Western science, and represented considerable progress in the science of meteorology.[20] In it Alhazen discussed atmospheric refraction and gave the first correct definition of twilight. He showed that twilight begins when the sun is 19° below the horizon.[21] Using this result, Alhazen attempted to measure the height of the atmosphere. In a very complicated geometric demonstration, drawing often upon Euclid, he arrived at a maximum height of 52,000 "passuum"—about forty-nine miles.[22]

Increasingly in the eleventh and twelfth centuries, Christian scholarship was stimulated by the circulation of Latin translations of Hindu and Greek texts preserved in the Western centers of

Moslem learning, like Palermo and Toledo. One of the earliest translators was the English monk, Adelard of Bath (ca. 1120). Little is known about him except that he studied in Spain, traveled extensively throughout the Levant, and is credited with one of the first Latin translations of Euclid's *Elements.*[23]

Adelard was quite interested in meteorological speculation. Those which appeared in his "Quaestiones Naturales" are novel because they are not just paraphrases of earlier Greek theories. For example, in his discussions of winds Adelard applied a theory which has as its basis the conjecture that the forms of all things were the cause of passive effects. Thus, air, which itself was not in a state of movement, produced movement which was the wind: "Wind then is air in a state of motion, and dense enough to have propulsive force; for I think that wind is a species of air." He then considered such questions as why winds traveled around the earth and how they gained their tremendous power.[24]

Among his other meteorological speculations is a discussion of thunder and lightning.[25] Again, his theory is different from any previously posed. He explained that thunder was due to the breaking of ice colliding in clouds; in the summer it was caused by the melting of the colliding ice. As for lightning, Adelard observed that in all violent collisions of bodies, the lightest thing in them was the first to be separated from them. The fire-like aether in the air was the lightest substance in the air, and the violent collisions of ice in clouds forced this aether out of the air, causing the lightning.[26]

In the thirteenth century the authority of Aristotle was completely reestablished in meteorology.[27] His systematically developed theories were generally superior to others then existing and consequently were immediately adopted. For the next four centuries these theories enjoyed almost uncontested acceptance, and meteorological work simply consisted of commentaries on them. The number of these produced between the beginning of the thirteenth and the middle of the seventeenth centuries was prodigious. It has been estimated that well over 156 commentaries on Aristotle's *Meteorologica* appeared before 1650.[28] In the thirteenth century some of the better known commentators were St. Thomas Aquinas (1225–1274), the Franciscan monk Bartholomaeus Anglicus (fl. 1220–1250), Robert Grosseteste (1168–1253) in England, Thomas of Cantimpre (ca. 1200–1270) in Belgium,

Vincent de Beauvais (ca. 1190–1264) in France, and the German Albertus Magnus (1206–1280).

Medieval research was very largely research in the library. This was especially true in meteorology. In any problem, whenever the choice arose between going out to nature or back to the books, the cloistered scholar retreated to his books. Argument from authority outweighed experimental evidence. Before meteorology developed further, this supremacy had to be broken. This was a slow process, but a definite beginning was made by the most prominent English scholar of the thirteenth century, Roger Bacon (1214–1294).

This great Franciscan scientist, who is often said to have heralded the dawn of modern science in Europe,[29] energetically advocated experimentation and the mathematical approach in all scientific study, including meteorology.[30] In his *Opus Majus*, Bacon followed Aristotle's version of an atmosphere composed of water, air, and fire concentrically surrounding the earth. His work in optics made it clear to him that the atmosphere was of varying density.[31] He gave geometric proof of Aristotle's hypothesis that the global shape of air must be spherical within and without.[32] Bacon also discussed the topic of climates, referring to the work of Ptolemy.[33] He noted that Ptolemy's general climatic zones had to be adjusted to account for different topographies, such as mountains high enough to "exclude the cold from the north" and thus effect the climate of the area.[34] Bacon also joined the army of commentators on Aristotle's *Meteorologica* with his work *In Meteora*. It was, however, his insistence on the importance of experimentation and observation in science rather than on the authority of ancient writers which was Bacon's prime contribution to the development of meteorology. He made the first step towards freeing meteorology from the bonds of Aristotelian theories, chains which were finally broken in the seventeenth century.

The 400 years after Bacon were the lull before the seventeenth-century storm. By the middle of the sixteenth century, meteorology had developed along two divergent lines: the theoretical or pure science based on Aristotle's *Meteorologica*, and the applied pseudo-science of weather prognostications carried on by the astrologers. Weather prediction by natural signs became very popular;[35] astrometeorology enjoyed the protection

Girolamo Cardona
(1501-1576)

René Descartes
(1596–1650)

of the church and of princes. That the astrologers who published weather prognostications included noted cultural and scientific figures like Johann Muller (Regiomontanus) (1436–1476), Leonard Digges (ca. 1550), and Johann Kepler (1571–1630), helped to establish accreditation for this pseudo-science. Typical examples of sixteenth-century astrological weather forecasts appeared in Digges' *A Prognostication of Right Good Effect* (1555):[36]

> The Sunne in the Horizon, or ryfynge, clear and bright, fheweth a pleafant day: gut thinlie overcaft with a cloude, betokeneth foule weather.
>
> If thyck clowdes refemblying flockes, or rather great heapes of wall, be gatherid in many places, they fhewe rayne.
>
> Thundres in the morning fignifie wynd: about noone, rayn: in the evenygh great tempeft.

By no means were all of the scientists between the thirteenth and seventeenth centuries convinced of the validity of astrological weather prognostications. One doubter was the noted mathematician and commentator on the *Meteorologica,* Nicole Oresme (1323–1382). Oresme, who was one of the first to recognize the great problems involved in weather prediction, had little regard for the astrologers of his time. He believed that weather forecasting was possible, but thought that the proper rules for it were still unknown[37]—a belief shared by most early twentieth-century meteorologists. Despite dissenters like Oresme, however, the influence of astrology persisted until the beginning of the eighteenth century, long after it had lost all pretence to scientific consideration.

As we have seen, Aristotle's *Meteorologica* dominated meteorology from antiquity to the seventeenth century and retarded its development as a science. Theoretical speculation by such scientists as Robert Recorde (1510–1558) and John Dee (1527–1608)[38] largely consisted of rehashing Aristotelian theories. The sixteenth century, however, saw a gradual breaking away from this influence. The crack in Aristotle's authority, made a few centuries earlier by Bacon, was widened by the noted algebraist Girolamo Cardano (1501–1576).

In his noted treatise *De Subtilitate* (1550), Cardano devoted considerable attention to meteorological speculation.[39] His discussions of such atmospheric phenomena as winds, clouds, rain, and lightning reflected the strong Aristotelian influence still prevalent in Western Europe. In his discussion of air, however, Cardano took issue with several basic theories of Aristotle. For one thing, he maintained, there were three, not four, basic elements: earth, air, and water.[40] Fire was not a basic element since it required food to exist, moved about, and produced nothing. Like Aristotle, Cardano separated air into two parts but in a different way. He explained there were two kinds of air—"free air," which destroyed inanimate things and preserved animate things, and "inclosed air," which had the opposite effect.[41]

If Cardano's *De Subtilitate* marked the beginning of the end of the 2,000-year reign of the *Meteorologica* as the prime authority for theoretical meteorology, the end itself occurred in the seventeenth century, the "insurgent century" of science.[42] Scientists no longer accepted without question the classical theories of Aristotle and his contemporaries. As they became unshackled from classical authority, these scientists more and more turned to the experimental method in their search for truth. This new scientific spirit was reflected in the works of such seventeenth-century giants as Galileo, Huygens, and Pascal. As will be seen, all three made significant contributions to the development of meteorology. The one man, however, who most typified both the end of the era of speculation and the birth of modern meteorology was the "first modern philosopher," René Descartes (1596–1650).

In 1637, Descartes published his renowned book *Discours de la Méthode*, in which he expounded his philosophy of scientific method. The basis of this method was four-fold:

1. Never to accept anything as true unless one clearly knew it to be such.
2. To divide every difficult problem into small parts, and to solve the problem by attacking these parts.
3. To always proceed from the simple to the complex, seeking everywhere relationships.
4. To be as complete and thorough as possible in scientific investigations allowing no prejudice in judgment.

In an appendix to this work, "Les Météores,"[43] Descartes applied the above principles to his discussions on meteorology.

Like Aristotle, Descartes employed the deductive approach in his discussions.[44] In "Les Météores," he attempted to explain the nature and cause of all kinds of weather phenomena by showing that they are based on certain general principles of Nature which hitherto had not been adequately explained. Descartes began by discussing the nature of terrestrial bodies, and the vapors which arise from them; then he explained the formation of clouds and winds, along with the manner in which the clouds dissolve to form rain, hail, and snow. After classifying the causes of tempests, thunder, and lightning, he closed with a discussion of the effects of light which cause rainbows and other luminous phenomena in the heavens.

In these discussions, Descartes first presented the hypothesis that air, water, and other terrestrial bodies were all composed of small particles, and between these were interstices filled with every "subtile matter."[45] He supposed that the particles of water were long, smooth, and slippery, "like little eels, which, though they join and twist around each other, do not, for all that, ever knot or hook together in such a way that they cannot easily be separated." The particles of hard substances were interlaced and joined together, and had irregular shapes. If the particles were smaller and less interwoven, they could be more agitated by the particles of the subtile matter which were always in motion, and thus they formed either air or oils. All Descartes' meteorological discussions were based on this hypothesis.

Although Descartes rejected Aristotle's meteorological theories, some of his own reflected an Aristotelian influence. According to Descartes, winds were caused by several things: vapors which were drawn upwards by the sun from terrestrial substances such as clouds dissolving into vapors, and also clouds which fell and drove away the air which was under them. If a cloud suddenly fell on another cloud below it, the result was thunder; lightning was due to the presence of inflammable exhalations between the two clouds.[46] Descartes' explanations of clouds, rain, etc., are quite modern. He explained that clouds were composed of drops of water or small pieces of ice. These drops were formed by the coalescence of small particles of vapor, and were round unless their shape was altered by the wind. When their size became so large that the air would not hold them, they fell as rain, as snow if the air was not warm enough to melt them, or as hail if after

DISCOVRS

DE LA METHODE

POVR BIEN CONDVIRE SA RAISON, & chercher la verité dans les Sciences.

PLVS

LA DIOPTRIQVE, LES METEORES; LA MECHANIQVE, ET LA MVSIQVE, Qui sont des essais de cette METHODE,

PAR RENE' DESCARTES.

Auec des Remarques & des éclaircissemens necessaires.

A PARIS,
Chez CHARLES ANGOT, ruë saint Iacques, au Lion d'Or.

M. DC. LXVIII.

AVEC PRIVILEGE DV ROY.

Title Page to Descartes'
Discours de la Méthode

being melted they met a colder wind which froze them.[47]

The purpose of "Les Météores" was to demonstrate the superiority of Descartes' method applied to meteorology over those previously proposed (in particular Aristotle's).[48] Like his predecessors, however, he suffered from a lack of any scientific means of studying the atmosphere other than by visual observation. Thus, he also had to resort to a deductive method in his explanation of weather phenomena which, due to his erroneous scheme of the universe, was not successful.[49]

Meteorology, however, owes a great deal to René Descartes. The Cartesian coordinate system is as important in meteorology as it is in other sciences. Like Roger Bacon, he stressed the importance of mathematics to meteorology and to all physical science—"My Physics is nothing but Geometry."[50] "Les Météores" was instrumental in the establishment of meteorology as a legitimate branch of physics.[51] And it was Descartes' enthusiastic interest in meteorology which proved a major catalyst for its rebirth as a true science in the seventeenth century.

References

1. Theophrastus, *Enquiry into Plants and Minor Works on Odours and Weather Signs,* trans. Sir Arthur Hort (London: William Heinemann Ltd., 1948), 2: 395.

2. *Ibid.*

3. Proclus, *The Commentaries of Proclus on the Timaeus of Plato,* trans. Thomas Taylor (London: the author, 1820), pp. 100–102.

4. For example, see J.I. Craig, "Meteorological Conditions Controlling the Nile Flood," *Quart. Jour. of the Roy. Meteor. Soc.* 35 (1909): 141–143; and E.W. Bliss, "The Nile Flood and World Weather," *Quart. Jour. of the Roy. Meteor. Soc.* 53 (1927) 41–43.

5. Secundus Plinius, *Pliny: Natural History,* trans. H. Rackham (London: William Heinemann Ltd., 1938), 1: 229.

6. Claudius Ptolemaeus, *Ptolemy's Tetrabibles,* trans. F.E. Robbins (Cambridge: Harvard University Press, 1940), p. 215.

7. S.K. Heninger, Jr., *A Handbook of Renaissance Meteorology* (Durham: Duke University Press, 1960), p. 217.

8. Julius Hann, *Handbuch Der Klimatologie* (Stuttgart: Verlag Von J. Engelhorm, 1883), p. 58.

9. For a full account of the Tower of the Winds, see Stuart and Revett, *Antiquities of Athens* (London, 1972). Also see Thompson, D'Arcy, "The Greek Winds," *Classical Review* 32 (1918): 49.

10. Sir William Napier Shaw, *Manual of Meteorology* (Cambridge: The University Press, 1926) 1: 81.

11. See Seneca, *Quaestiones Naturales,* trans. John Clarke (London: Macmillan and Co., Ltd., 1910).

12. See Plinius, *op. cit.,* pp. 245–289.

13. Cicely M. Botley, "A Founder of English Meteorology," *Quart. Jour. of the Roy. Meteor. Soc.* 61 (1935): 346.

14. Venerabilis Bede, *The Complete Works of Venerable Bede: In the Original Latin,* ed. J.A. Giles (London: Whittaker and Co., 1843), Vol. VI.

15. *Ibid.,* p. 115.

16. For a discussion of Isidore's role in medieval science, see Floyd S. Lear, "St. Isidore and Medieval Science," *The Rice Institute Pamphlet* 33 (1936): 75–105.

17. Heninger, *op. cit.,* p. 721.

18. L. Dufour, "Les grandes époques de l'histoire de la météorologia," *Ciel et Terre* 59 (1943): 357.

19. Sarton, *op. cit.,* p. 721.

20. *Ibid.*

21. Alhazen, *Opticae Thesaurus,* ed. Federico Risnero, Basileae, per Episcopios (1572), pp. 286–287.

22. *Ibid.,* p. 288.

23. Carl B. Boyer, *A History of Mathematics* (New York: John Wiley & Sons, Inc., 1968), pp. 276–277.

24. Adelard of Bath, "Quaestiones Naturales," *Dodi Ve Nechdi* (Uncle and Nephew), trans. Hermann Gollancy (London: Oxford University Press, 1920), pp. 145–148.

25. For a historical discussion of the theories on the cause of thunder and lightning, see H. Howard Frisinger, "Early Theories on the Cause of Thunder and Lightning," *Bulletin of the American Meteorological Society* 46, No. 12 (1965): 785–787.

26. Adelard of Bath, *op. cit.,* p. 149.

27. A.C. Crombie. *Augustine to Galileo—The History of Science* (Cambridge: Harvard University Press, 1953), p. 70.
28. Dufour, *op. cit.*, p. 357.
29. Sir Oliver Lodge, *Pioneers of Science* (New York: Dover, 1960), p. 9.
30. Roger Bacon, *The Opus Majus of Roger Bacon,* trans. Robert B. Binke (Philadelphia: University of Pennsylvania Press, 1928), 1: 116.
31. *Ibid.*, p. 153.
32. *Ibid.*, p. 178.
33. *Ibid.*, pp. 154-158.
34. *Ibid.*, p. 155.
35. See Curt F. Bühler, "Sixteenth-Century Prognostications," *Isis* 33 (1941-42): 609-620.
36. Leonard Digges, *A Prognostication of Right Good Effect* (London: 1555).
37. Lynn Thorndike, *A History of Magic and Experimental Science* (New York: Columbia University Press, 1934), 3: 416-417.
38. Robert Recorde, *The Castle of Knowledge* (London: 1556), pp. 6, 64, 91; John Dee, *De nubium, solis, lunae ac reliquorum planetarum, imo, ipsius stelliferi coeli, ab intimo terrae centro distantus mutuisqui intervallis et eorundem omnium magnitudine* (London: 1551). This is a work on clouds.
39. Girolamo Cardano, *De Subtilitate,* 3rd ed. (Basel: 1569).
40. *Ibid.*, p. 375.
41. *Ibid.*, p. 396.
42. Charles Singer, *A Short History of Scientific Ideas to 1900* (London: Oxford University Press, 1959), p. 218.
43. For a discussion of "Les Météores," see F. Wootton, "The Physical Works of Descartes," *Science Progress* 21 (1927): 457-478.
44. To see how Descartes applied his method in the natural sciences, see Hyman Stock, *The Method of Descartes in the Natural Sciences* (New York: The Marion Press, 1931).
45. René Descartes, *Discours De le Méthod* . . . (Paris: Charles Angot, 1668), pp. 189-195.
46. *Ibid.*, pp. 253-256.
47. *Ibid.*, pp. 227-236.
48. Etienne H. Gilson, "Météores cartesians et météores scholastiques," *Études de Philosophie Mediévale* (Paris: J. Vrin, 1921): 247-286.

49. In this connection, see Rufus Suter, "Science Without Experiment: A Study of Descartes," *Scientific Monthly* 58 (1944): 265–268.

50. Carl B. Boyer, *The Rainbow* (New York: Thomas Yoseloff, 1959), p. 201.

51. A. Wolf, *A History of Science Technology, and Philosophy in the 16th and 17th Centuries,* 2nd ed. (London: George Allen and Unwin Ltd., 1950), p. 306.

The Dawn of Scientific Meteorology

Galileo
(1564–1642)

Chapter Four

The Thermometer

At the close of the sixteenth century, meteorology, which had been based on the speculations of natural philosophers such as Aristotle, was at an impasse. It had become increasingly evident that these speculations were often erroneous and inadequate and that a greater knowledge of the atmosphere was an imperative step to any improvement in the science. Descartes himself recognized the pressing need for a greater knowledge of the atmosphere and became a pioneer in its acquisition. To obtain this knowledge, meteorological instruments were needed which had not yet been developed. Fortunately, this lack was soon to be rectified.

For it was in the seventeenth century that what are probably the three most basic instruments in meteorology were first developed; the thermometer, the barometer, and the hygrometer. Of the first two, the noted English meteorologist Sir Napier Shaw states, "The invention of the barometer and thermometer marks the dawn of the real study of the physics of the atmosphere, the quantitative study by which alone we are enabled to form any true conception of its structure."[1] This chapter is devoted to a discussion of the early development of the thermometer.

Although not discovered until the early 1600's, the basic principle behind it, the expansion of air by heat and its contraction by cold, was known much earlier—at least as far back as the third century B.C. and the experiments of Philo of Byzantium.[2] In his *De Ingeniis Spiritualibus* (On Pressure Engines), Philo described an apparatus (*Fig. 4.1*) which illustrated the expansion and contraction property of air:[3]

> One takes a leaden globe of moderate size, the inside of which is empty and roomy. It must neither be too thin that it cannot easily burst, nor too heavy, but quite dry so that the experiment may succeed. Through an aperture in the top is passed a bent siphon reaching nearly to the bottom. The other end of this siphon is passed into a vessel filled with water, also reaching nearly to the bottom, so that water may the more easily flow out. (a) is the globe, (b) the siphon, and (g) the vessel. I assert, when the globe is placed in the sun and becomes warm, some of the air enclosed in the tube will pass

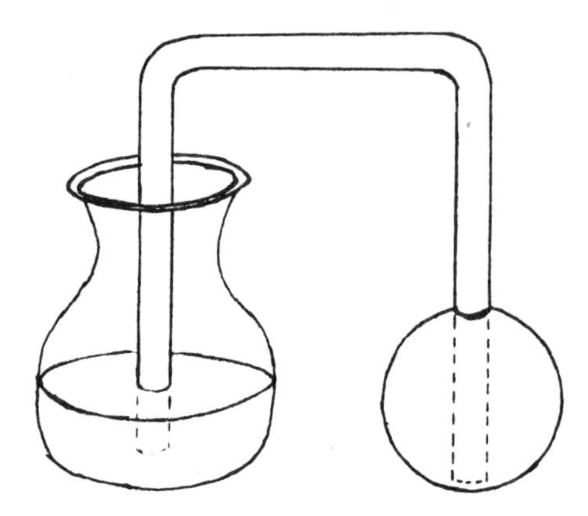

Philo's Apparatus for His Experiments with Air

Fig. 4.1

> out. This will be seen, since the air flows out of the tube into the water, setting it in motion and producing air-bubbles, one after the other. If the globe is placed in the shadow or any other place where the sun does not penetrate, then the water will rise through the tube flowing into the globe. If the globe is again placed in the sun the water will return to the vessel, . . . The same effect is produced if one heats the globe with fire or pours hot water over it. . . .

A later and better known writer, Hero of Alexandria, constructed a similar apparatus, as well as a primitive steam-engine which illustrated the principle of a jet engine.

Although Philo and Hero both demonstrated the expansion and contraction property of air, neither seemed to recognize the great applicability of this property for measuring heat. For this, credit generally goes to the famous Italian physicist and astronomer, Galileo Galilei.[4] The son of an impoverished Florentine nobleman, Galileo was born at Pisa, February 18, 1564, the day of Michelangelo's death. Although to please his father, Galileo entered the University of Pisa in 1581 to study medicine, his love of mathematics overcame parental objections: For the next thirty years he devoted his attention largely to applied mathematics and general science. During this period, he was professor of mathematics at the Universities of Pisa (1589-1591) and Padua (1592-1610) where he established a European reputation as a scientist and inventor.[5]

It was at Padua that Galileo invented the first thermometer. The exact date is not certain, but Galileo might have designed it as early as 1593. At any rate, it has been clearly established that

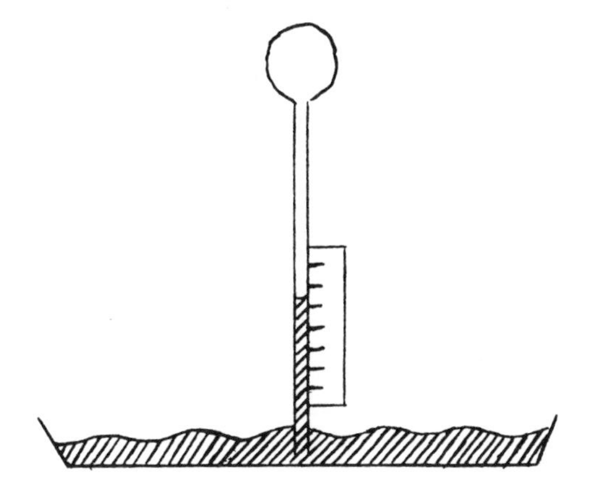

Galileo's Thermometer
Fig. 4.2

Galileo was demonstrating a thermometer in his famous public lectures shortly after the beginning of the seventeenth century. In a letter written by Father Castelli to Monsignor Cesarini dated September 20, 1638 in which he stated that Galileo had used his thermometer in public lectures thirty-five years previously, Father Castelli provided the following description of Galileo's thermometer (*Fig. 4.2*):[6]

> Galileo took a glass vessel about the size of a hen's egg, fitted to a tube the width of a straw and about two spans long; he heated the glass bulb in his hands and turned the glass upside down so that the tube dipped in water held in another vessel; as soon as the ball cooled down the water rose in the tube to the height of a span above the level in the vessel; this instrument he used to investigate degrees of heat and cold.

Galileo used his thermometers to determine the relative temperatures of different places and of the same place at different times and seasons. His method of graduating the stem of his thermometers is not known and was undoubtedly arbitrary. That he cited "degrees" in his measurements, however, indicates a higher type of instrument than some thermoscopes of even later periods.[7] It must be remembered, incidentally, that the inverted air thermometers of Galileo were subject to changes in atmospheric pressure, and thus no two of them were comparable. The development of sealed thermometers dependent upon the expansion of liquids and independent of atmospheric pressure did not occur for another fifty years.[8]

The seventeenth century was not very old before the news of

Galileo's new instrument for measuring temperatures spread throughout Europe. Soon many other scientists were constructing their own thermometers, each attempting to improve on the others. It is not feasible to attempt a description of each one of the numerous early thermometers.[9] To obtain a general picture of their great variations, however, a few will be briefly considered. One of these early innovators was a French theologian, Father Marin Mersenne.

The early seventeenth century preceded the founding of academies of science and periodical literature. Communication between European scientists was maintained by personal visits and by correspondence. Probably the most noted and active intermediary between seventeenth-century scientists was Father Mersenne, who was in constant communication with such eminent men as Galileo, Descartes, Huygens, and Roberval. Undoubtedly, Mersenne played an important role in the rapid dissemination of the news of Galileo's thermometer. Mersenne also originated the still popular custom of stimulating scientific efforts by proposing prize questions.[10]

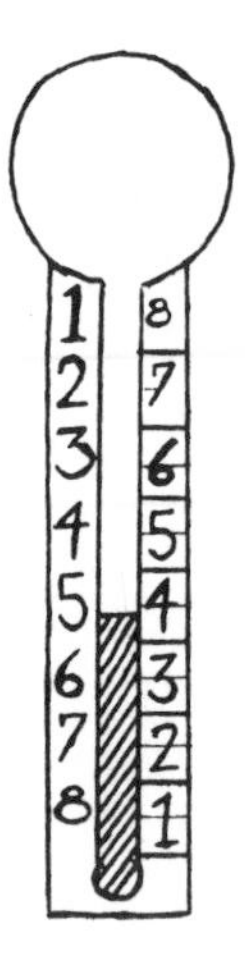

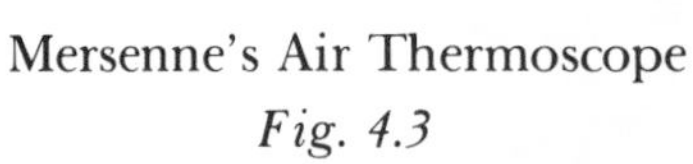

Mersenne's Air Thermoscope
Fig. 4.3

Ferdinand's Thermometer
Fig. 4.4

In addition to acting as an intermediary, Mersenne would often repeat the experiments of others. Thus, he devised a modification of the air thermoscope intended to increase its sensitivity to temperature change (*Fig. 4.3*).[11] In a work published

Mersenne
(1588–1648)

at Paris in 1644, Mersenne described the instrument as a narrow tube with a large bulb at one end and a small one at the other, the latter being pierced by a minute hole. The larger bulb was gently warmed while the smaller was first submerged in water, when the heating of the larger bulb was stopped, a few drops of the liquid rose in the graduated tube to form a short column of water that served as an indicator of temperature change.

Other early innovators working on the thermometer were the Frenchman Jean Fey and the Belgian John Baptist van Helmont. It was in Italy, however, where the next important development occurred under the direction of Ferdinand II, Grand Duke of Tuscany, a renowned scientific patron. Around 1641, Ferdinand, who did considerable research in physical science, turned his attention to improving the air thermometer. He constructed a typical model, filled it to a certain height, and then sealed it hermetically by melting the glass tip. This closed instrument was then graduated by degrees marked on the stem (*Fig. 4.4*).[12] By sealing the top, Ferdinand had constructed what was probably the first thermometer independent of atmospheric pressure, a vital step in the development of an instrument applicable for accurate temperature investigations.[13] Ferdinand used this and other thermometers in the atmospheric temperature investigations performed by the famous Accademia del Cimento which he helped to establish in 1657. The important contribution of this society to the creation of a network of meteorological observation stations throughout Europe will be discussed in chapter seven.

Undoubtedly, one of the most unusual instruments in the evolution of the thermometer was constructed in 1660–1662 by the German physicist Otto Von Guericke (1602–1686). This particular instrument was nearly twenty feet long (*Fig. 4.5*) and consisted of a large copper glove (painted blue with gold stars) joined to a long copper tube one inch wide. This tube was bent upon itself to form a very narrow U in which a given amount of alcohol was placed. The shorter arm of the U was open at the top and on the liquid floated a tiny inverted brass foil cup to which a cord was attached that passed around a wheel hung upon the underside of the globe. On the other end of the cord was a little figure of an angel pointing to the scale on the tube. A valve at one side of the large copper sphere was used to withdraw enough air by means of an air pump to adjust the height of the alcohol. This immense thermometer was attached to the shady side of a house and was renowned for its ability to show "the coldest and hottest weather throughout an entire year."[14]

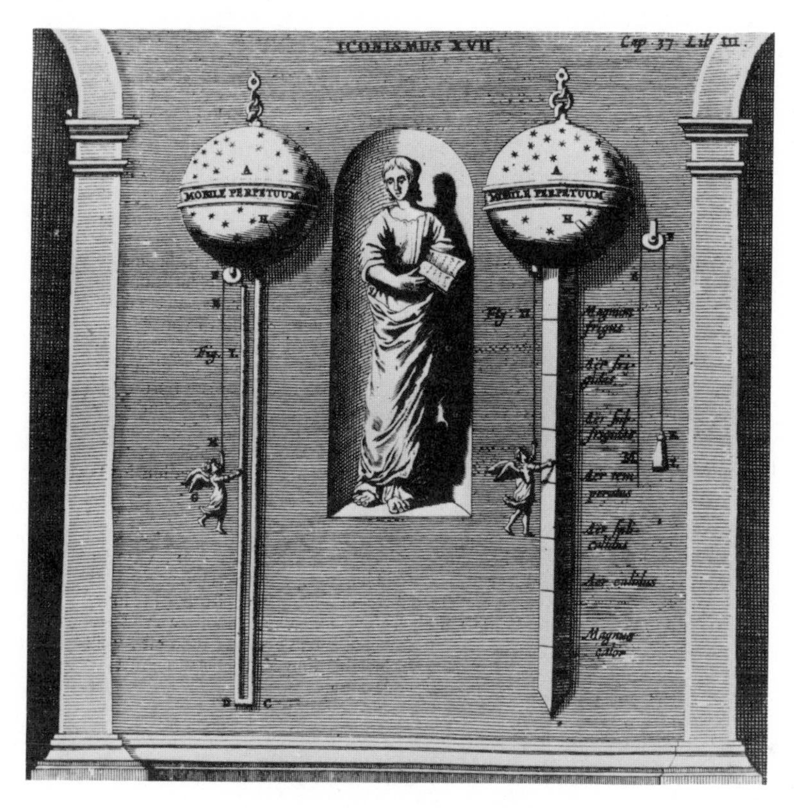

Guericke's Thermometer
Fig. 4.5

Ferdinand's invention of a thermometer that was independent of pressure opened the door to systematic study of surface temperature variations. Before such a study could proceed very far, however, the scales on the thermometers had to be standardized. One of the first to recognize the need for a standard scale for thermometers was the noted English physicist Robert Boyle (1627–1691), whose work in chemistry and physics well qualified him for thermometrical studies. He devoted considerable effort to constructing sealed thermometers and recognized the pressing need for a standard scale to permit the comparison of effects shown by different models:[15]

Robert Boyle
(1627–1691)

> We are greatly at a loss for a standard whereby to measure cold. The common instruments show us no more than the relative coldness of the air, but leave us in the dark as to the positive degree thereof; whence we cannot communicate the idea of any such degree to another person. For not only the several differences of this quality have no names assigned them, but our sense of feeling cannot therein be depended upon; and the thermometers are such very variable things that it seems morally impossible from them to settle such a measure of coldness as we have of time, distance, weight, etc.

Boyle endeavored to overcome this difficulty. Believing that the melting point of ice varied with latitude, he suggested the use of aniseed oil to obtain a fixed point by placing the oil around the bulb of an alcohol thermometer, allowing the oil to freeze, and then marking the height of the alcohol "when the oil begins to curdle."[16] He attempted to compute the absolute expansion of alcohol and to divide the scale into ten thousandths, or some other numerical part of the total expansion.

The fixed point of Boyle's did not prove very successful, and it remained for a colleague, Robert Hooke, to make the next important advance in the standardization of thermometric scales. Hooke's scientific work was extensive and for the most part bore the mark of genius. Not the least of his work was a tremendous contribution to the emerging science of meteorology, especially in instrumentation. Professor Andrade has gone so far as to say that Hooke "invented all the meteorological instruments."[17] Although this claim is not literally correct, it is true that Hooke helped significantly to develop several meteorological instruments, including the thermometer.

In October 1664, Hooke published his noted *Micrographia*, in which he described his work with sealed thermometers which he had "by several tryals, at last brought to a great certainty and tenderness: for I have made some with stems above four foot long, in which the expanding Liquor would so far vary, as to be very neer the very top in the heat of Summer, and prety neer the bottom at the coldest time of the winter." He filled the thermometers with "best rectified spirit of wine highly ting'd with the lovely colour of cochineal."[18] An important advance was made when, in his graduation of the stem, he placed zero at the point where the liquid stood when the bulb was placed in freezing distilled water, marking the divisions above and below "according the degrees of

Christian Huygens
(1629–1695)

expansion or contraction of the liquor in proportion to the bulb it had when it indur'd the freezing cold." According to his contemporary, Edmund Halley, Hooke once exhibited at Gresham College a combination of barometer and thermometer with the stem of the thermometer graduated from −70 to +130 (freezing point of water = zero) (*Fig. 4.6*).[19] In January 1665, Hooke was instructed to produce a thermometer for the Royal Society that would serve as a standard for observation.[20] His thermometers were used by the Society for several years.

While Hooke was the first to take the freezing point of water as a fixed point on the thermometric scale, the Dutch mathematician Christian Huygens was probably the first to suggest two fixed points, the second point being that of boiling water.[21] Huygens made this proposal in a letter written to Robert Moray, dated January 2, 1665:[22]

> It would be well to have a universal and determinate standard for heat and cold, securing a definite proportion between the capacity of the bulb and the tube, and then taking for the commencement the degree of cold at which water begins to freeze, or better the temperature of boiling water, so that without sending a thermometer to a distance, one could communicate the degrees of heat or of cold found in experiments and record them for the use of posterity.

It is not clear in this letter, however, whether these two fixed points were to be used together, or as alternatives.

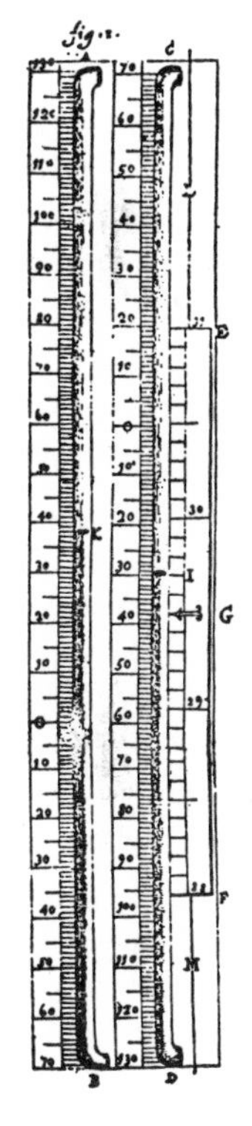

Fig. 4.6 Hooke's Barometer-Thermometer

Nearly thirty years later in 1694, the use of both the freezing point and the boiling point of water as fixed points in thermometric scales was again proposed by Carlo Renaldini, a former member of the famous Academia del Cimento, and professor of mathematics in Padua.[23] He also suggested dividing the space between these two fixed points into twelve equal parts. Unfortunately, because many seventeenth-century scientists doubted the constancy of the two fixed temperature points, Renaldini's important proposal was not conclusively adopted until the eighteenth century.[24]

By the close of the seventeenth century, the problem of obtaining a satisfactory thermometric scale was being considered by many leading scientists, including Sir Isaac Newton

MICROGRAPHIA:

OR SOME

Physiological Descriptions

OF

MINUTE BODIES

MADE BY

MAGNIFYING GLASSES.

WITH

OBSERVATIONS and INQUIRIES thereupon.

By *R. HOOKE*, Fellow of the ROYAL SOCIETY.

Non possis oculo quantum contendere Linceus,
Non tamen idcirco contemnas Lippus inungi. Horat. Ep. lib. 1.

LONDON, Printed by *Jo. Martyn*, and *Ja. Allestry*, Printers to the ROYAL SOCIETY, and are to be sold at their Shop at the *Bell* in S. *Paul's* Church-yard. M DC LX V.

Title Page to Robert Hooke's *Micrographia*

(1642–1727). In 1701, Newton anonymously published a paper in the *Philosophical Transactions* on the results of his thermometric researches.[25] His thermometer was three feet long with a bulb two inches in diameter. The dilating liquid used was not alcohol, but "lintseed oil." For fixed points in the scale, Newton chose the temperature of melting snow and of the human body, dividing the interval into twelve equal parts. Hew drew up a scale of heat covering the range from the freezing point of water to the heat of a coal fire (*Fig. 4.7*).

Equal Parts of Heat	Degrees of Heat	Phenomena
0	0	The heat of air in winter, when the water begins to freeze.
12	1	Greatest heat the thermometer received on the surface of a human body, as also that of a bird hatching her eggs.
24	2	The heat of melting wax.
48	3	The lowest heat at which equal parts of tin and bismuth melt.
96	4	The lowest heat at which lead melts.
192	5	The heat of a small coal fire not urged by bellows.

Newton's Scale of Heat Table
Fig. 4.7

Assuming that the rate of cooling of a hot body was proportional to the temperature of that body, Newton was able to estimate higher temperatures, such as "red heat," by observing the times needed by the bodies to cool to temperatures measurable on his thermometer. Later called "Newton's Law of Cooling," this assumed proportion stated that:[26]

> For the heat which hot iron, in a determinate time, communicates to cold bodies near it, that is, the heat which the iron loses in a certain time, is as the whole heat of the iron;[27] and therefore, if equal times of cooling be taken, the degrees of heat will be in geometrical proportion, and therefore easily found by the tables of logarithms.

Sir Isaac Newton
(1642–1727)

Although his proposal for a standard scale for thermometers was not very satisfactory, Newton's discovery of the law of cooling in solid bodies and his observation of the constancy of temperature in fusion and ebullition was of major importance for the development of thermometers.[28]

The impetus provided by Hooke, Newton and others, along with the rapidly increasing demands of physics, medicine and meteorology led to a surge of activity in eighteenth-century thermometry. Many articles on the subject appeared in the journals of prominent scientific societies, the Royal Society of London and L'Academie Royale Des Sciences of Paris. Three scientists in particular were foremost in the field: Gabriel Fahrenheit, René de Réaumur, and Anders Celsius.

Fahrenheit worked on thermometry as early as 1706. He first employed only alcohol, but later used mercury. There is considerable doubt exactly where and when the first mercury thermometer appeared. One claim is that the French mathematician Ismael Boullian (1605–1694) was the first to use mercury in a thermometer, while others maintain that as early as 1657 a mercury thermometer was being used by members of the Accademia del Cimento.[29] It is indisputable, however, that in 1714 Fahrenheit constructed the first mercury thermometers with reliable scales,[30] which then became famous throughout Europe.

Although his publications are few and brief, they reveal an original thinker. In one of these papers he presented an interesting account of his thermometers.[31]

> The thermometers constructed by me are chiefly of two kinds, one is filled with alcohol and the other with mercury. Their length varies with the use to which they are put, but all the instruments have this in common: the degrees of their scales agree with one another and their variations are between fixed limits. The scales of thermometers used for meteorological observations being below with 0° and go to 96°. The division of the scale depends upon three fixed points which are obtained in the following manner: The first point below at the beginning of the scale, was found by a mixture of ice, water and sal-ammoniac, or also sea-salt; when a thermometer is put in such a mixture the liquid falls until it reaches a point designated as zero. This experiment succeeds better in winter than in summer. The second point is obtained when water and ice are mixed without the salts; when a thermometer is put into this mixture the liquid stands at 32°, and this I call the commencement of freezing, for still water

> becomes coated with a film of ice in winter when the liquid in the thermometer reaches that point. The third point is at 96°; the alcohol expands to this height when the thermometer is placed in the mouth, or the arm-pit, of a healthy man and held there until it acquires the temperature of the body.

When Fahrenheit made thermometers for higher temperatures, the scale was simply lengthened by adding more spaces of equal length. Interestingly, the boiling point of water was not one of Fahrenheit's fixed points—a division on his scale marked 212 just happened to coincide with it.

Fahrenheit's work was an important step in the development of accurate thermometers not only because he popularized the use of mercury (his thermometers were adopted throughout England and Holland) but also because he demonstrated that other liquids, like water, had fixed boiling points which varied with a change in atmospheric pressure.

René Antoine Ferchault de Réaumur (1683–1757) is probably best known for his thermometric research. In 1730, he published a paper in the *Memoirs of the Academy of Science* entitled: "Regles Pour Construire Des Thermometres Doni Les Degre's Soient Comparables, ..."[32] This was followed by a long paper on the same subject published in the *Memoirs* the following year.[33] Like so many Frenchmen, Réaumur completely ignored Fahrenheit's work, and rejected the use of mercury because of its small expansion coefficient in favor of alcohol. He marked the freezing point of water zero, its boiling point 80, and divided the intervening space into eighty parts. Despite its ingenuity, his thermometer was severely criticized for ignoring the influence of air pressure on the boiling points of liquids, a property that Fahrenheit had already discovered. Moreover, with bulbs three to four inches in diameter, they were too large and awkward for general use. For these and other reasons they proved unsatisfactory and today Réaumur is chiefly remembered for his scale, a version of which is still employed on the Continent.

The Swedish astronomer, Anders Celsius, is the third scientist prominent in the standardization of thermometric scales in the first half of the eighteenth century. In a paper read before the Swedish Academy of Sciences in 1742, he proposed a new scale for thermometers.[34] In his centigrade thermometer Celsius used two fixed points: the temperature of boiling water, which he assigned the value of zero on his scale, and the temperature of melting ice, which he assigned the value of 100 (*Fig. 4.8*). This centesimal graduation proved much more satisfactory than most of the previous scales and gradually became that most prominently employed in scientific work. It should be noted, however, that the present centigrade thermometer has its scale reversed from that of Celsius' original thermometer, with 100 degrees for the boiling point of water, and zero for its freezing point. This inversion was apparently first introduced in 1743 by Jean Pierre Christin of Lyons.[35]

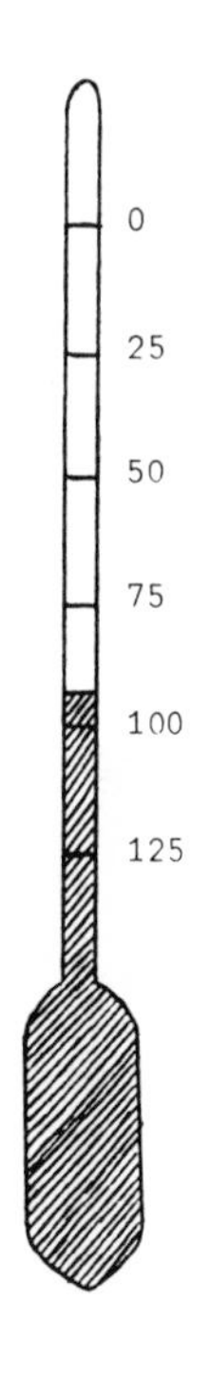

Fig. 4.8 Celsius Thermometer

Although many improvements were still to be made (e.g., standardization of thermometric scales—in 1779 Johann H. Lambert listed nineteen in current use[36]), by the close of the eighteenth century the thermometer had reached a sufficient stage of development to allow for satisfactory use in scientific investigations.[37]

References

1. Sir William Napier Shaw, *Manual of Meteorology* (Cambridge: The University Press, 1926), 1:115.

2. A.C. Crombie, *Augustine to Galileo: The History of Science* (Cambridge: Harvard University Press, 1953), p. 346.

3. Gustav Hellmann, "The Dawn of Meteorology," *Quart. Jour. of the Roy. Meteor. Soc.* 34 (1908): 226-227.

4. There are three additional serious candidates for the honor of having "invented the thermometer," Santorio Santorre, Robert Fludd, and Cornelius Drebbel. However, most historians credit the "invention" to Galileo. See W.E. Knowles Middleton. *A History of the Thermometer and Its Uses in Meteorology* (Baltimore: The Johns Hopkins Press, 1966), pp. 4-23.

5. See. R. E. Gibson, "Our Heritage from Galileo Galilei," *Science* (Sept. 1964): 1271-76.

6. Florian Cajori, *A History of Physics* (New York: The Macmillan Co., 1906), p. 90.

7. For further discussion on Galileo's work in the field of temperature, see Rolin Wavre, "Galileé et le probleme du temps," *Gesnerus* I (1943): 25-34.

8. A little etymology might be of some interest at this point. According to Henry C. Bolton, the word "thermoscope" first appears in print in the treatise "Sphaera mundi, seu cosmographia demonstrative" written in 1617 by

Giuseppe Bianconi and printed in 1620 at Bologna. The word "thermometer" is first found in the Jesuit Father Jean Leurechon's book *La Recreation Mathematique,* printed in 1624, in which he states "thermometer, instrument for measuring degrees of heat and cold that are in the air." Henry C. Bolton, *Evolution of the Thermometer 1592-1743* (Easton, Pa: The Chemical Publishing Co., 1900), p. 11.

9. For a thorough discussion on the many early thermometers, see the above work by H.C. Bolton or the excellent book by W.E. Knowles Middleton cited in footnote 4.

10. Bolton, *op. cit.,* p. 29.

11. The air thermometer had become quite common by the second quarter of the seventeenth century.

12. Gustav Hellmann, "Die Altesten Quecksilber Thermometer," *Meteorologische Zeitschrift* 14 (1897): 31-32.

13. Charles Singer, *A Short History of Science* (Oxford: The Clarendon Press, 1959), p. 348.

14. Bolton, *op. cit.,* p. 47.

15. E. Gerland, "Zur Geschichte des Thermometers," *Zeitschrift Fur Instrumentenkunde* 13 (1893): 341-342.

16. Bolton, *op. cit.,* p. 43.

17. E.N. Andrade, "Robert Hooke, 1635-1703," *Nature* 171, No. 4348 (1953): 366.

18. Robert Hooke, *Micrographia* (London, 1665), p. 38.

19. Edmund Halley, "On the Expansion and Contraction of Fluids by Heat and Cold, in order to Ascertain the Divisions of the Thermometer, and to make that Instrument, in all places, without Adjusting by a Standard," *Philosophical Transactions of the Royal Society of London* 17 (1693): 650-657.

20. Louise D. Patterson, "The Royal Society's Standard Thermometer 1663-1709," *Isis* 44 (1953): 52.

21. Harvey A. Zinszer, "Meteorological Mileposts," *Scientific Monthly* 58 (1944): 261.

22. Christian Huygens, *Oeuvres completes de Christiaan Huygens publiees par la Societe Hollandaise des Sciences* (The Hague: 1893), 5:188.

23. W.E. Knowles Middleton, *A History of the Thermometer and Its Uses in Meteorology* (Baltimore: The Johns Hopkins Press, 1966), p. 55.

24. Cajori, *op. cit.,* p. 92.

25. Isaac Newton, "A Scale of the Degrees of Heat," *Phil. Trans.* 22 (1701): 824-829.

26. *Ibid.,* p. 827.

27. By whole heat, Newton means the temperature of the iron.

28. A. Wolf, *A History of Science Technology, and Philosophy in the 16th and 17th Centuries*, 2nd ed. (London: George Allen & Unwin Ltd., 1950), p. 89.

29. G. Hellmann, "Die Altesten Quecksilber Thermometer," *op. cit.*, p. 31.

30. Bolton, *op. cit.*, p. 66.

31. Daniel G. Fahrenheit, "Experimenta et observationes de congelatione aquae in vacuo factae," *Phil. Trans.* 33 (1724): 78-89.

32. René A. Réaumur, "Regles Pour Construire Des Thermometres Doni Les Degre's Soient Comparables, . . . ," *Memoires De L'Academie Royal Des Sciences* (1730): 452-507.

33. Réaumur, "Second Memoire Sur La Construction Des Thermometres, Dont Les Degres sont Comparable . . . ," *Mem. Acad. Roy. Scie.* (1731): 250-296.

34. Anders Celsius, "Observationer om tvenne bestandiga grader pa en thermometer," *Vetenskap. Akad.*, Handl, 4 (1742): 197-205.

35. Middleton, *op. cit.*, pp. 101-105.

36. Cajori, *op. cit.*, p. 111.

37. For a thorough discussion of the more recent developments in thermometry, see Middleton, *op. cit.*, pp. 115-238.

Evangelista Torricelli
(1608–1647)

Chapter Five

The Barometer

The invention of this vital instrument which measures the atmospheric pressure or "weight" was another product of seventeenth-century investigations. Before discussing its early development, however, it might be appropriate to consider the basic concept of the barometer—the idea that air has weight.

Although the concept of air as a "heavy" body was not conclusively demonstrated until the sixteenth century, conjectures on the weightiness of the atmosphere occurred much earlier. As far back as the fourth century B.C., Aristotle had proved that air had body by showing that a vessel must be emptied of air before it could be filled with water.[1] Aristotle was also one of the first to suspect that air was heavy. To prove his suspicion, he took a leather bag and weighed it when it was "empty" of air (i.e., when the bag was pressed flat). Next he weighed the bag when it was "full" of air. To his disappointment, Aristotle found no difference, and concluded that air was weightless.[2]

The question, nonetheless, remained unsolved and even the great Galileo was puzzled. In 1615, he believed that air had no weight.[3] Later, however, he described an experiment in the *Discorsi e dimonstranzioni mathematiche* (1638) which seemed to prove that air did have weight. In this experiment Galileo used a syringe to force into a previously weighed and properly valved bottle an amount of air which at ordinary pressure would have been of two or three times its volume, and determined with precision an actual increase of weight. He then allowed the excess air to escape without loss through a tube which led to another bottle completely filled with water, and so constructed that the water displaced by the entering air could be caught in a suitable weighed vessel. Lastly, he determined the increase in weight of the second bottle.[4]

Part of Galileo's trouble was perhaps his confusion between air weight and air pressure, two separate ideas. This distinction often led to problems in the thinking of many seventeenth-century scientists. By the middle of the century, however, studies on the weight-of-air problem led to a momentous development in the history of the barometer, the "Torricellian Experiment."

The invention of the barometer is generally ascribed to Evangelista Torricelli (1608–1647), an Italian mathematician and disciple of Galileo.[5] He probably obtained the idea of his experiment both from Galileo's theory that no heavy body ascended without another equally heavy body descending in counterbalance to it, and his suggestion that the force counterbalancing the ascending liquid in a vacuum pump was the "secret and invisible pressure of the air."[6] Following this lead, Torricelli in 1643 at Florence had one of his own pupils, Vincenzo Viviani, conduct the famous "Torricellian Experiment" on the weight of air, in which the first barometer was constructed.

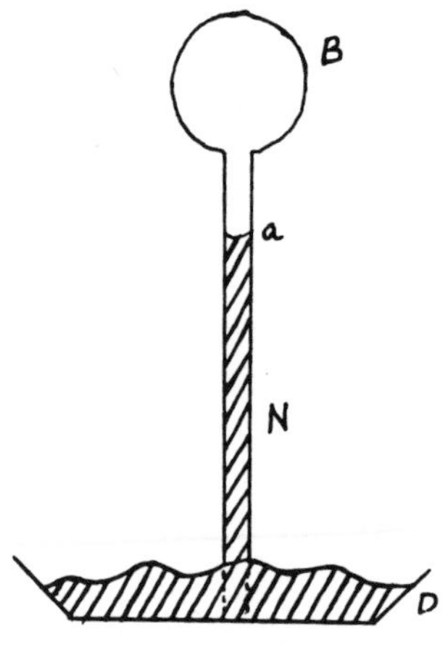

Torricelli's Barometer
Fig. 5.1

A description of Torricelli's experiment and his mercury barometer was given in a letter from him to Michelangelo Ricci on June 11, 1644. The barometer consisted of a glass bulb (B) with a neck (N) "two cubits" long. (A cubit varies in length, but is usually between seventeen and twenty-one inches.) The tube was filled with mercury and inverted in a basis (D) containing more mercury (*Fig. 5.1*). The mercury in the glass bulb and neck would fall to the level (a) and oscillate there with changes in atmospheric pressure. (Torricelli first used water as the liquid in his barometric model and had to use a cumbersome sixty-foot glass tube.[7] Later, using the much heavier mercury, he was able to reduce its height to about thirty-two inches.

Toricelli, however, ran into unexpected trouble:[8]

> I must add that my principal intention—which was to determine with the instrument when the air was thicker and heavier and when it was more rarefied and light—has not been fulfilled; for the level (a) changes from another cause (which I never would have believed), namely, on account of heat and cold: and changes very appreciably.

Thus, Torricelli's instrument not only acted like a barometer, but also like Galileo's thermometer.

There has been a considerable amount of controversy over the part that Descartes played in the invention of the barometer.

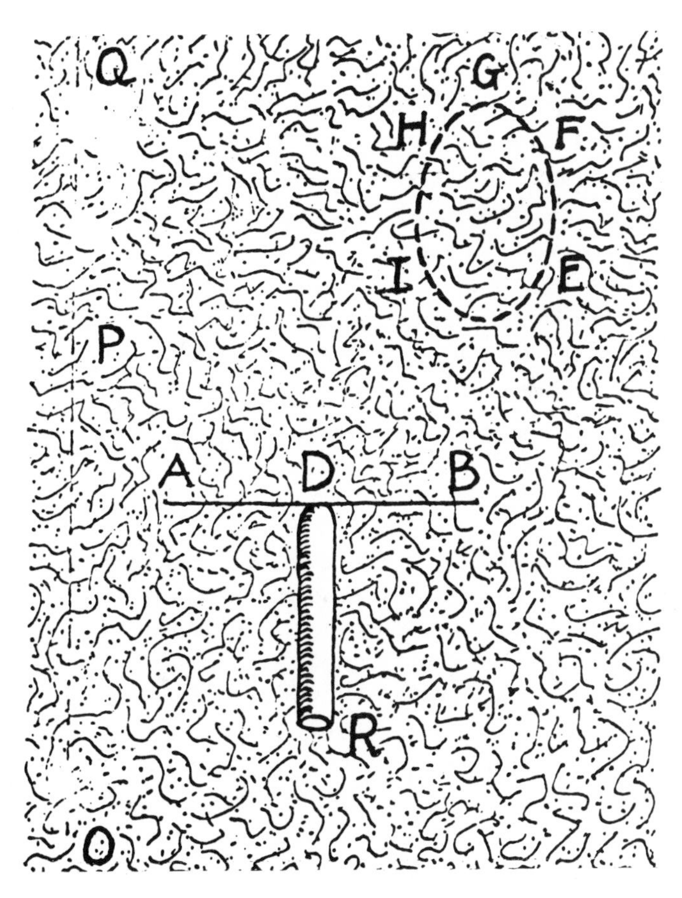

Descartes' Theory of Vortices
Fig. 5.2

On June 2, 1631, he wrote a letter to one of his pupils, Reneri, in which the subjects of atmospheric pressure and vacuum were discussed.[9] In this letter Descartes used wool as an analogue to air and made strong use of his theory of vortices.[10] He explained that weight is not commonly felt in the air when it is pushed upwards because the wool (air) at point E (*Fig. 5.2*) is raised towards F; that which is at F goes in a circle towards GHI and comes back to E, and thus there is no more weight than would be felt in turning a wheel. Next, he considered the example of a tube DR closed at the end D, and filled with mercury. As the mercury descends down the tube DR, the wool (air) at R moves toward 0, and that which is at 0 goes towards P and Q, etc., until all the wool (air) in the line OPQ is raised. Since the tube is closed at the top, no wool may enter it; so it takes some force to raise all the wool in the line OPQ. The force thus needed to separate the mercury from the top of the tube is just that force which is required to raise all the air from that point to above the clouds. Descartes never admitted the existence of a vacuum, so he supposed that the space in the tube left by the descending mercury was filled by the aether above the air, which came down and completed the circle of movement.[11]

The historian E. Gerland claimed that in the above letter Descartes had discussed the idea of the barometer well before the "Torricellian Experiment."[12] There is, however, no indication in this letter that Descartes realized a basic barometric concept, that if the tube DR were long enough only part of the mercury would come out. This is not to say that Descartes deserves no credit in the invention of the barometer. In a letter to Mersenne on December 13, 1647, Descartes wrote:[13]

> But, so that we may also know if changes of weather and of location make any difference to it, I am sending you a paper scale two and a half feet long, in which the third and fourth inches above two feet are divided into lines; and I am keeping an exactly similar one here, so that we may see whether our observations agree.

Hence, if a barometer must be an instrument with a scale, then Descartes was the first to make a barometer.[14]

There is another possible claimant to the honor of developing the first barometer, Gasparo Berti, about whom very little is known. He had a considerable reputation as a mathematician and astronomer, and it is possible that he performed a barometric experiment as early as 1639.[15] It was the "Torricellian Experiment," however, which caused the greatest agitation in the scientific community of Western Europe.

Torricelli never published an account of his barometric research, perhaps because he was too deeply absorbed in mathematical investigations on the cycloid.[16] He did, however, extensively describe his experiments in two letters of 1644, to his friend Michaelangelo A. Ricci in Rome. Later the same year, Ricci wrote a letter describing the Torricellian experiment to Pere Marin Mersenne in Paris,[17] who, with his extensive correspondence, spread the news throughout Europe.

It was nearly twenty years before any further significant advances were made in the construction of the barometer. Typical of this period was the barometer of the famous Accademia del Cimento, constructed about 1657. In 1665, however, there began a hundred-year period of intense activity in barometric development which encompassed the efforts of some of the era's leading scientists.[18]

Again, Robert Hooke was one of the outstanding contributors, with his invention of the "wheel barometer" in 1665.[19] This

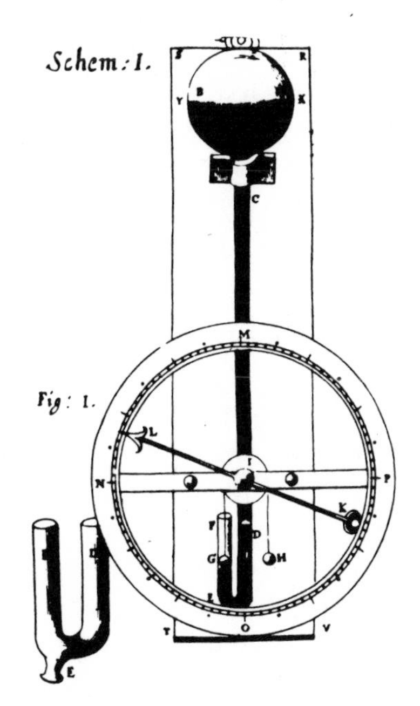

Hooke's Wheel Barometer
Fig. 5.3

still popular instrument consisted of a siphon mercury barometer, a circular scale and pulley arrangement (*Fig. 5.3*). A line from a float (G) in the short branch of the siphon mercury barometer was passed over a pulley carrying the index (I), and attached on the other side to a counterpoise (H): The index moving over a circular dial indicated thereon the barometric variations. Hooke continued to work on its improvement and invented other types of barometers, including a self-registering one.[20]

The word "barometer" itself was first used about this time, apparently by Robert Boyle.[21] Like his colleague Hooke, Boyle was also interested in both the thermometer and the barometer. He used the barometer in many of his investigations of atmospheric phenomena, including the question of the existence of a vacuum, about which he never seemed to be able to make up his mind.[22] In 1669, Boyle published a manuscript entitled *Continuation of New Experiments* which included a description of perhaps the first completely portable barometer, a siphon, that could be considered an instrument.[23] The Royal Society was impressed by its portability and resolved on June 4, 1668 that[24]

> the society being put in mind to give order for the making of portable baroscopes, by Mr. Boyle, to be sent into several parts of the world, the operator was ordered to attend Mr. Boyle, to receive his directions for filling them aright; and that being done, to make some of them forthwith, to be sent not only into the most distant places of England, but likewise by sea into the East and West Indies, and other parts, particularly to the English plantations, as Bermudas, Jamaica, Barbados, Virginia, and New England; and to Tangier, Moscow, St. Helena, the Cape of Good Hope, and Scanderoom; . . .

It is perhaps just as well that this grandiose plan was never executed, since it is very doubtful that Boyle's barometers would have survived the intended trips.

One result of the attempts to improve the sensitivity of the barometer to pressure changes was the invention of the two-liquid barometer. Descartes was apparently the first to construct one (*Fig. 5.4*) in which the bottom liquid was mercury and the top liquid was water.[25] The idea was to increase the sensitivity of the barometer to changes of pressure by employing liquids of lower specific gravity such as water or spirit of wine in conjunction with mercury. The magnification of such a barometer is close to the ratio of the specific gravities of the two liquids. Water, however, was a poor choice for the second liquid since the water vapor in the upper tube made the instrument very sensitive to temperature.

Robert Hooke in 1668 constructed and demonstrated to the Royal Society his version of the two-liquid barometer (*Fig. 5.5*)[26] which was a definite improvement over Descartes'. Although it is very doubtful that he was aware of Descartes' instrument, Hooke also used mercury and water. Note that readings were taken from the water surface at D.

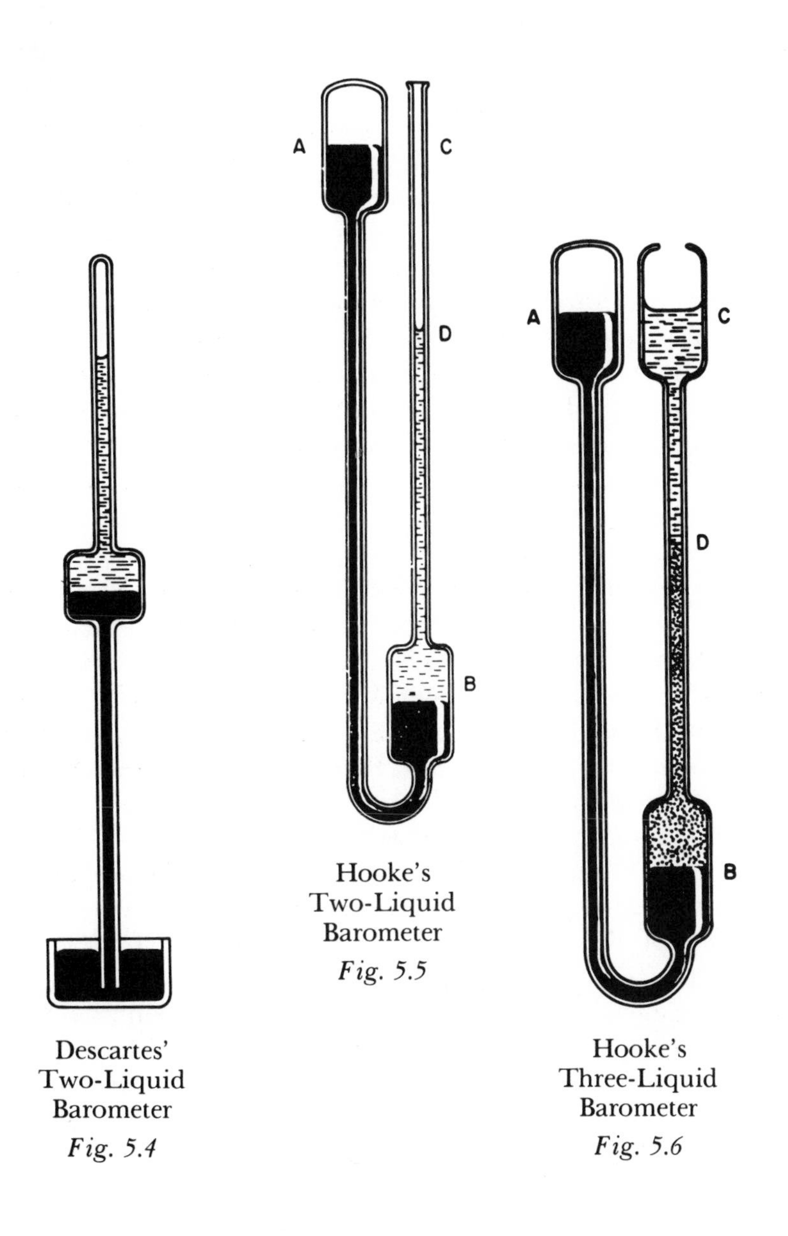

Descartes' Two-Liquid Barometer *Fig. 5.4*

Hooke's Two-Liquid Barometer *Fig. 5.5*

Hooke's Three-Liquid Barometer *Fig. 5.6*

Huygens was also attempting to improve the sensitivity of the barometer. Following Descartes' suggestions,[27] he reconstructed both Descartes' and Hooke's form of the two-liquid barometer in 1672.[28]

A major defect of Hooke's two-liquid barometer was that the tube containing the lighter liquid became dirty in the region through which the liquid moved. Both Hooke and Huygens were aware of this problem, but it was Hooke who proposed an effective solution to the problem. In 1685, Hooke described to the Royal Society a three-liquid barometer (*Fig. 5.6*) which resembled the two-liquid barometer with the addition at the top of the open tube of a reservoir C having the same diameter as the bulbs A and B. The tube BC now contained two immiscible liquids, for example, alcohol and turpentine with the indicator of the atmospheric pressure being the common surface D of these two liquids.

The two-liquid barometer of the type invented by Hooke and Huygens contained another disadvantage. When the atmospheric pressure increased, the barometric readings decreased. It was not until the latter half of the nineteenth century that the two-liquid barometer was modified so that it would register "right-side-up."[29] This instrument, nevertheless, became very popular in Western Europe and was one of the favorite barometers for two centuries.

Several other models were developed during this period in an attempt to improve and enlarge their scales. One variation was the diagonal barometer. This instrument, whose invention is generally credited to Sir Samuel Morland around 1688,[30] consisted of a simple modification of the "Torricellian" tube (*Fig. 5.7*). The tube was bent at the top so that the change in mercury levels was substantially increased. Another was the "square" or L-shaped barometer which is usually ascribed to Johann Bernoulli (1667–1748),[31] one of the eight members of the famous family of mathematicians. Bernoulli's barometer contained two tubes, the diameter of the vertical tube being greater than that of the horizontal (*Fig. 5.8*). It was based on the concept that if the vertical tube was k times as wide as the horizontal tube, the magnification of the change in mercury levels would be k^2. This instrument, however, had several serious disadvantages which appeared later during attempts to calibrate it and make temperature corrections. As a result, it enjoyed little popularity.

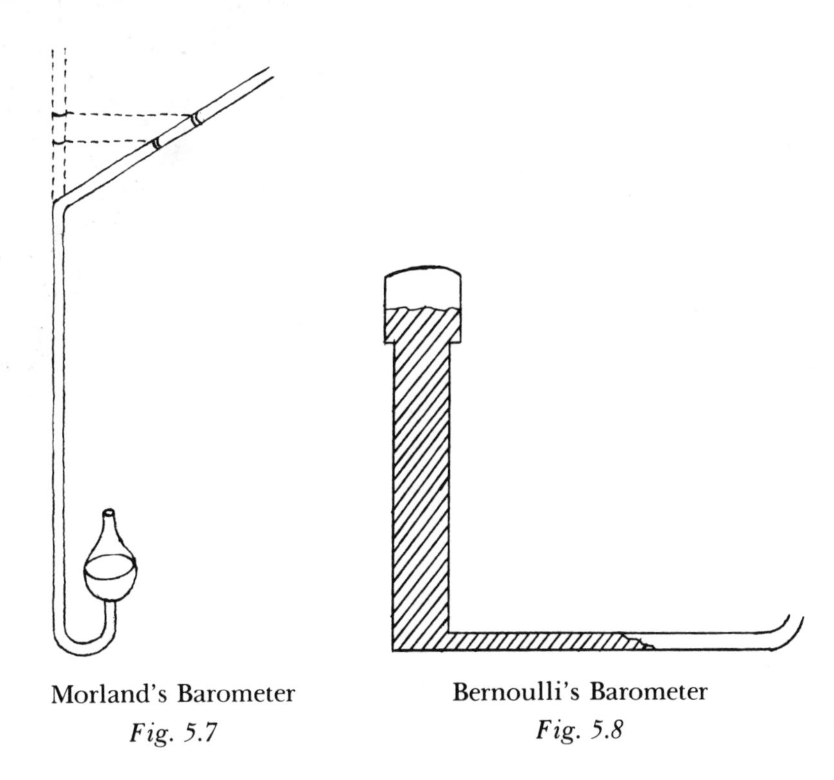

Morland's Barometer
Fig. 5.7

Bernoulli's Barometer
Fig. 5.8

The early usefulness of barometers was seriously hampered by their great delicacy. This was especially true in the frequent attempts to use them to measure heights, especially of mountains. The search for portability, therefore, was one of the important factors in their early development. One obvious possible solution to the problem was to fill the barometer with mercury at the site of the proposed observation. Pierre Bouguer (1698–1758) devised a method of accomplishing this,[32] but his idea proved unsatisfactory and attention soon turned to the construction of a more compact and sturdy instrument. A forerunner in these efforts was the celebrated Swiss meteorologist, Jean Andre De Luc (1727–1817).

In 1749, De Luc started to improve the barometer. As with the thermometer, De Luc was highly dissatisfied with the barometers then in use. Like his predecessors, he realized that there were incessant fluctuations in the height of the mercury column not caused by pressure changes. He spent much time studying the cause of these fluctuations, particularly those caused by temperature changes, and calculating the temperature corrections to be applied to barometric readings. He also constructed the first good portable siphon barometer,[33] which consisted of a J-shaped tube in two parts connected through a cock in the shorter arm of the tube. When the longer arm was closed, the shorter arm was open at the top. Whenever the instrument was to be transported any distance, the longer arm was completely filled with mercury and sealed by closing the cock, and whatever mercury remained in the shorter tube was drained out. Thus, the danger of damage from violent agitation of the mercury in the longer arm was avoided. To obtain temperature corrections for the barometric readings, a thermometer was mounted on the base. The entire instrument was enclosed in a box and carried upside down like a quiver. At the observation site it was mounted vertically, with the aid of a plumb line, upon a tripod.[34]

De Luc's work with barometers triggered a heated debate over the relative merits of the two major types—the cistern and the siphon—which was to last over a hundred years. Although the siphon barometer had been used by Pascal and others as early as the late seventeenth century, nearly all barometers with any pretense to accuracy prior to 1770 were of the cistern type. When De Luc developed the first fairly accurate portable siphon barometer and expressed his strong preference for the siphon barometer over the cistern barometer, the great battle of siphon versus cistern was joined. In short order, the famed physicist and pioneer in electrical science, Henry Cavendish (1731–1810), replied to De Luc, giving his full support to the cistern barometer.[35] The battle was waged throughout the nineteenth century; the siphon barometer, backed by such noted scientists as Gay-Lussac, had the upper hand during the first half of the century, but the cistern barometer gradually triumphed. Thus, at least in the English-speaking countries, the cistern barometer has been the dominant twentieth-century type.

Another eighteenth-century barometer was the aneroid which dates back to about 1700 and originated with the great Gottfried Wilhelm Leibniz (1646–1716) who carried on a voluminous correspondence with other European mathematicians, in particular Johann Bernoulli. In one of these letters (June 7, 1698), he pondered over "a small closed bellows which would be compressed and dilated by itself as the weight of the air increases or decreases."[36] Four years later, in another letter to Johann

Bernoulli (February 3, 1702),[37] he became much more definite: "... about a portable barometer which could be put in the pocket, like a watch; but it is without mercury, whose office the bellows performs, which the weight of the air tries to compress against the resistance of the steel spring." According to Hellmann, Bernoulli replied that all the materials Leizniz suggested for a bellows would be adversely affected by humidity.[38] On June 24, 1702 Leibniz answered, "I should like to use a metallic bellows, in which the folds will be furnished with strips of steel. In this way the effects you fear will be nullified."[39]

It is clear that Leibniz had the ideas of the basic concept and construction of the aneroid barometer well in mind, but apparently was unable to find anyone capable of its construction. In fact, nearly one and a half centuries were to elapse before such an instrument was successfully constructed. Leibniz can thus be considered the inventor of the aneroid barometer only in that he was the first to propose it.

After the work of De Luc, the major barometric developments occurred in the nineteenth century—standardization of barometric measurements, corrections for temperature, capillarity, etc. By the close of the eighteenth century, however, the barometer was indisputably a reliable meteorological instrument actively employed by meteorologists throughout the western world.

References

1. A.C. Crombie, *Augustine to Galileo—The History of Science* (Cambridge: Harvard University Press, 1953), p. 238.

2. L.F. Kaemtz, *A Complete Course in Meteorology* trans. C.V. Walker (London: Hippolyte Bailliere, 1845), p. 232.

3. Galileo, *Le opere*, ed. nat. (Florence: 1884), 4:167.

4. Galileo, *Dialogues Concerning Two New Sciences*, trans. Henry Crew and Alfonso DeSalvio (New York: The Macmillan Company, 1914), pp. 79-83.

5. Harvey A. Zinszer, "Meteorological Mileposts," *Scientific Monthly* 58 (1944): 261.

6. E. Saul, *An Account of the Barometer* (London: 1730), pp. 4-5.

7. *Ibid.*, p. 5.

8. Blaise Pascal, *The Physical Treatise of Pascal*, trans. I.H.B. and A.G.H. Spiers (New York: Columbia University Press, 1937), p. 166.

9. René Descartes, *Ouevres de Descartes*, ed. Chas. Adam et Paul Tannery (Paris: 1897), 1:205-208.

10. According to Descartes, a vortex was a collection of particles of very subtle matter endowed with a rapid rotary motion around an axis which is also the axis of a sun or a planet. He attempted to account for the formation of the universe, and the movements of the bodies composing it, by a theory of vortices. For a concise discussion of this theory of vortices, see "Descartes," *Encyclopaedia Britannica*, 11th ed. (1911), 7:86.

11. The aether was supposed to be such a subtle matter that it could penetrate the pores of the tube.

12. E. Gerland, "Report of the International Meteorological Congress Held at Chicago, Ill., August 21-24, 1893," ed. O.L. Fassig, *U.S. Weather Bureau Bulletin*, No. 11, Part 3 (Washington, D.C.: 1896): 690.

13. Descartes, *op. cit.*, (1903 ed.), 5:99.

14. W.E.K. Middleton. *The History of the Barometer* (Baltimore: The Johns Hopkins Press, 1964), p. 46.

15. For a thorough discussion of what little is known concerning Berti's work, see W.E.K. Middleton, *The History of the Barometer* (Baltimore: The Johns Hopkins Press, 1964), pp. 10-18.

16. Florian Cajori, *A History of Physics* (New York: The Macmillan Company, 1906), p. 65.

17. *Ibid.*

18. For a complete and excellent discussion on the development of the barometer, see W.E.K. Middleton, *The History of the Barometer* (Baltimore: The Johns Hopkins Press, 1964).

19. Robert Hooke, *Micrographia* (London: 1665). The wheel barometer is described in the Preface of this work.

20. Hooke," "Self-registering instrument," *Trans. Roy. Soc. Edinburgh* 15 (1678): 503. See also William Ellis, "Brief Historical Account of the Barometer," *Quart. Jour. of the Roy. Meteor. Soc.* 12 (1886): 149.

21. Middleton, *op. cit.*, p. 71.

22. *Ibid.*, p. 75.

23. Robert Boyle, *Continuation of New Experiments* (1969) pp. 68-73.

24. *Phil. Trans.* 7 (1672): 5027-30.

25. Blaise Pascal, *Traitez de l'equilibre des liqueurs*, 2nd ed. (Paris: 1698), pp. 207-208.

26. Thomas Birch, *History of the Royal Society* (London: 1756), 2: 298.

27. Joachim D'Alencé, *Curieux traité de mathematique ou par le mouyen de trois instruments, a sauvoir, du barometre, du thermometre, du notiometre, du hygiometre . . .* , (Paris: 1713), pp. 19-20.

28. Christian Huygens, *Oeuvres Completes De Christiaan Huygens,* (La Haye: Martinus Nijhoff, 1897), 7:238-242.

29. Middleton, *op. cit.*, pp. 91-92.

30. *Ibid.*, pp. 110-111.

31. Johann Bernoulli's treatise, "Le baromètre in èquerre," is given in Jean De Luc, *Recherches sur Les Modifications de L'Atmosphere,* (Paris: 1784), 1:29-36. This type of instrument had previously been proposed by J. Dominic Cassini. See Middleton, *op. cit.*, pp. 115-116.

32. See the note by the Abbé de la Caille in Pierre Bouguer's posthumous *Traite d'Optique* (Paris: 1760), p. 323.

33. Jean De Luc, *Recherches sur les modifications de l'atmosphere,* 2nd ed., (Paris: 1784), 2:3-5.

34. A Wolf, *A History of Science, Technology, and Philosophy in the 18th Century* (New York: The Macmillan Co., 1939), p. 306.

35. *Phil. Trans.* 66 (1776): 381.

36. *Virorum Celeberr. Got. Gul. Leibnitti et Johan. Bernoulli Commercium Philosophicum et Mathematicum,* 2 vols. (Lausanne and Geneva, 1745), p. 368.

37. *Ibid.*, II, p. 70.

38. G. Hellmann, *Meteorol. Zeits.* 8 (1891): 158-159.

39. *Virorum Celeberr. . . . ,* *op. cit.*, 2:78.

Johann Lambert
(1728–1777)

Chapter Six

The Hygrometer and Other Meteorological Instruments

As was noted in Chapter I, the ancient Greeks had a fairly accurate grasp of the hydrological cycle—the ascent of water from the lakes, rivers, and oceans into the atmosphere, and its return again to the earth as precipitation. One problem with this cyclic concept was that at some stage in the process the water became invisible. To resolve this difficulty, Aristotle and other Greek natural philosophers assumed that some, but not all, of the ascending water turned into air.[1] The reason for the "hedge" on not having all of the rising water turning into air was that Aristotle realized that clouds consisted of drops of water.[2]

It was not until the seventeenth century that scientists influenced by Descartes began to accept the theory that water vapor was a distinct substance. Descartes held that all matter was composed of tiny, uniform particles distinguished from one another by their shape (see Chapter III), which for water took the form of long, smooth, eel-shaped particles easily separated.[3] He also noted that even when water became invisible, its particles maintained their shape, distinct from that of the particles of air.

This important advance, nonetheless, did not explain why water vapor stayed suspended in the air, and although it was the solution revealed by Daniel Bernoulli in the noted work *Hydrodynamica* (1738) which detailed his kinetic theory of gases,[4] it was generally ignored until the nineteenth century. The popular theory held by most eighteenth-century scientists was that somehow water dissolved in air much like solids dissolved in a liquid.[5]

These questions on the makeup and properties of water vapor, however, did not impede the attempts to invent an instrument capable of measuring the amount of water vapor in the atmosphere. In fact, the first invention of a humidity-measuring instrument occurred over a hundred years before the first thermometer. The man credited with this invention was the German, Cardinal Nicholas de Cusa (1401–1464), a fifteenth-century mathematician of some prominence. One of his works included a description of the earliest known hygrometer, perhaps the first meteorological instrument with a reacting substance:[6]

> If you suspend from one side of a large balance a large quantity of wool, and from the other side stones so that they weigh equally in dry air, then you will see that when the air inclines toward dampness, the weight of the wool increases, and when the air tends to dryness, it decreases.[7]

This type of hygrometer was still in use in the early part of the eighteenth century.[8]

During the next two hundred years little attention was given to improving this method of indicating humidity changes, although sometime around 1500 Leonardo da Vinci was reported to have constructed a mechanical moisture indicator.[9] In the second half of the seventeenth century, it was once again Robert Hooke who turned his many talents to the construction of an improved hygrometer. It had been observed that the pitch of the strings of musical instruments varied with the dampness of air.[10] In his *Micrographia,* Hooke used this property to discuss his attempts to construct a hygrometer employing catgut. He found greater success, however, using the "beard of a wild oat" as the reacting substance of his instrument.[11] The beard was to be firmly attached to the base of a well-ventilated box, passed through a hole in the center of a dial on the top of the box, and then fastened to a light index hand of black paper (*Fig. 6.1*). As the beard curled or uncurled according to the varying humidity, the index pointed over the various divisions on the dial.

Several other hygrometers were constructed in the latter part of the century. Santorio Santorre constructed three distinct types, the most popular being the string-hygrometer.[12] But the real development of the hygrometer as a scientific instrument did not begin until the eighteenth century and extensive investigations into the properties of hygrometry of the celebrated German mathematician Johann Heinrich Lambert (1728–1777).[13]

Although Lambert's meteorological investigations covered a wide range of topics from the barometer to the measurements of wind directions, he is best known as the first to suggest the name "Hygrometer."[14] His investigations in humidity measurements were compiled in 1774 in his "Suite de L'Essai d'Hygrometrie,"[15] which included a discussion of a catgut hygrometer constructed in 1768 (*Fig. 6.2*). The catgut would wreathe or unwreathe according to a decrease or increase in the moisture content of the

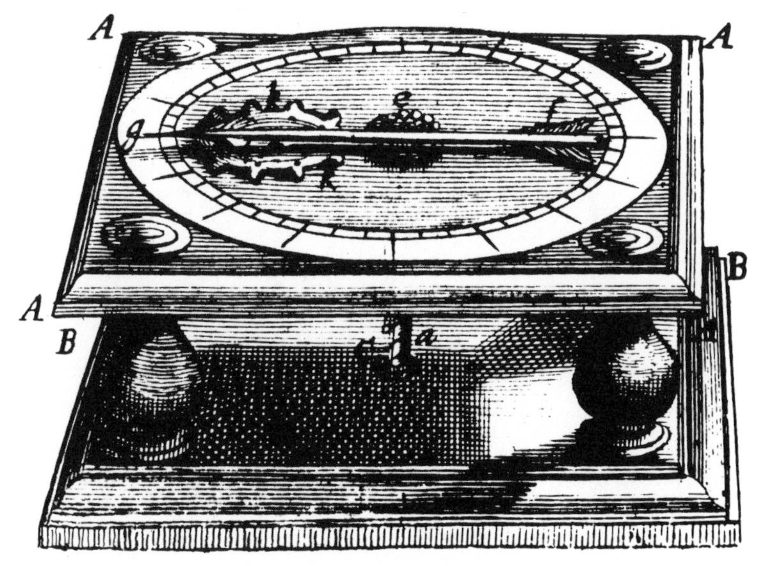

Hooke's Hygrometer
Fig. 6.1

surrounding air. Lambert performed many experiments with his hygrometer, always attempting to express his results mathematically.

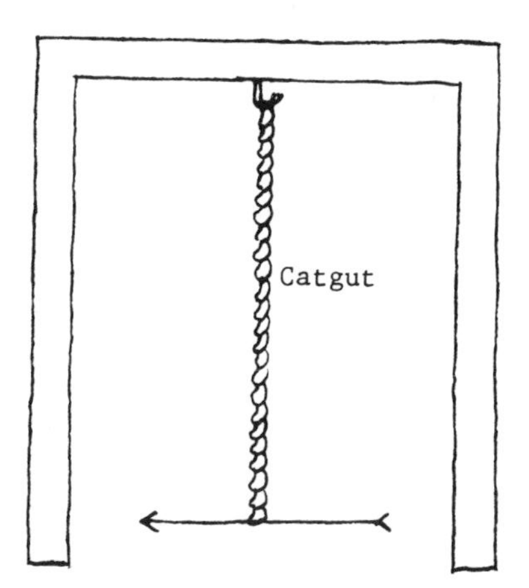

Lambert's Catgut Hygrometer
Fig. 6.2

He used the hygrometer to study the atmosphere in several meteorological investigations of monthly and annual variations in the atmospheric humidity in which he was extremely interested. He made a graph of the comparative daily hygrometer readings at three different cities, Berlin, Sagan, and Wittenberg. Lambert also arrived at an interesting correlation between humidity and temperature (*Fig. 6.3*).

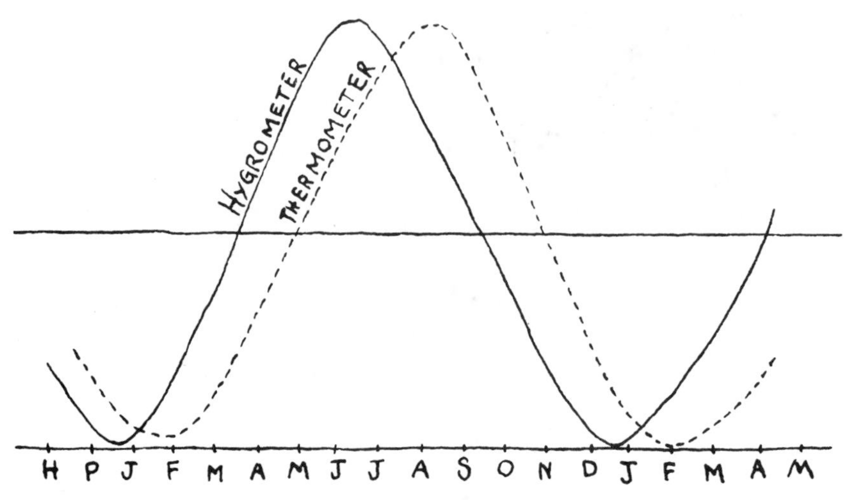

Lambert's Humidity-Temperature Correlation Graph
Fig. 6.3

Noting that the graph of the hygrometer preceded that of the thermometer by a greater extent in the spring and autumn than at other times of the year, Lambert concluded that the ordinates of both curves were functions of the longitude of the sun and could be expressed by the equation

$$y = A + B \sin r + C \cos r + D \sin 2r + E \cos 2r + \dots$$

$$(r = \text{longitude of the sun})$$

which was strongly convergent. He considered that this equation "representera assez bien les variations annuelles moyennes de l'humidité."[16]

The importance of this work lay not so much with Lambert's researches with the hygrometer as with the graphical presentation of his results. Lambert appears to have been the first to present meteorological data in graphs rather than in tables. It is Lambert's early realization of the great advantage of graphical representation for the analysis of meteorological data for which he should be most remembered in the history of meteorology.

Jean De Luc was also active in the development of the hygrometer. He described his hygrometers in a series of papers published in the *Philosophical Transactions* during the period 1773–1791. De Luc claimed that there were three essential requisites for a good hygrometer—a fixed point in the hygrometer scale, standardization of hygrometer scales, and equal differences in the pointer-reading to be produced by equal differences in humidity.[17] He considered that extreme humidity offered the only chance of establishing a fixed point and devoted a considerable amount of time to this problem.

In his first paper (1773), De Luc described the first hygrometer that was probably a truly scientific instrument.[18] It consisted of an ivory cylinder, about an inch long and about a quarter of an inch in diameter, which was closed at one end and attached at the other to a thermometer tube. The cylinder and part of the tube were filled with mercury. Thus, any variation in the size of the ivory cylinder produced by changes in the humidity of the air caused a substantial movement in the column of mercury in the base of the thermometer. In later experiments De Luc settled on strips of whalebone cut transversely to the fiber as the reacting substance which yielded the most uniform and satisfactory results. In subsequent papers he described his attempts to improve this hygrometer, experiments with the best methods for determining the fixed points of hygrometers, and his efforts to obtain a standard hygrometric scale. His hygrometers, the most noteworthy of the eighteenth century, utilized organic materials for the reacting substance of the instrument.[19]

Between 1780 and 1830 a great number of hygrometers were developed. One of the most prodigious writers in the field was the Swiss savant Horace Benedict de Saussure (1740–1799).

In 1783, De Saussure published the result of years of hygrometric research, the *Essais sur l'Hygrometrie*,[20] which consisted of four Essays. The first contained a detailed description of his famous hair hygrometer (*Fig. 6.4*). Earlier De Saussure had found that an ordinary human hair would vary one forty-second part of its length between the two extremes of complete dryness and saturation (he also found that this uniform action was materially impaired if the hair was not thoroughly freed from grease). His hygrometer was constructed to utilize this principle. While the lower end of the hair, *ab*, was connected to a clamp *b*, at the foot of the instrument; the other end was attached to *a*, a strip of foil wound on the horizontal cylinder *d* which, as it revolved, carried a pointer round a graduated dial. A counterpoise *g* suspended by a

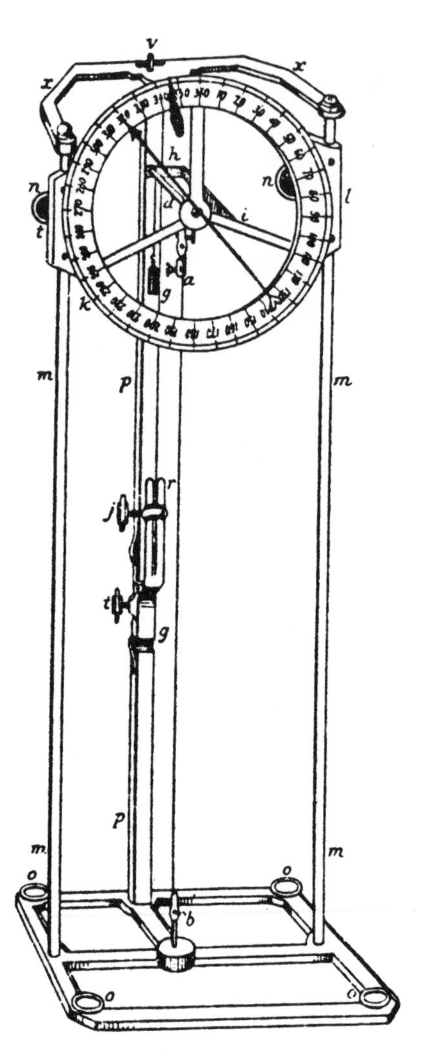

De Saussure's Hair Hygrometer
Fig. 6.4

silk thread, which was wound on the cylinder the opposite way to the foil, kept the hair taut. De Saussure found the above instrument too fragile for safe transport on his expeditions, and in the same Essay described a more compact and portable form of hair hygrometer.

In the second Essay, De Saussure discussed the general principles of hygrometry and in the last two dealt with evaporation and the application of his hygrometric results to the outstanding problems of meteorology. This work received wide acclaim throughout Europe. The famous and influential French naturalist, Georges Cuvier (1769–1832), regarded De Saussure's *Essays* as one of the greatest contributions made to science in the eighteenth century.[21] In his other researches De Saussure also showed that damp air was lighter than dry air at the same temperature and pressure.

During this same period a lively debate developed between the relative merits of De Saussure's hair hygrometers and De Luc's whalebone hygrometer with both authors active participants. Although De Luc's hygrometer had enjoyed a greater popularity, after 1820 the hair hygrometer was the most widely used hygroscopic hygrometer for serious meteorological work.[22]

In Italy a different type, the condensation hygrometer, had been developed in the late 1650's by the celebrated Accademia del Cimento under the leadership of the Grand Duke of Tuscany, Ferdinand II. He had

> . . . observed that when something iced was put in a glass the surrounding air seemed to change itself into water; whence he thought of filling a covered vessel with ice, and when this was done the air was converted into water. This accomplished, he cast about for an easier way of doing it, and this was to make the instrument D, which was a hollow glass vase coming to a fine point at one end; it was filled with crushed ice, and held in a wooden tripod stand, with a beaker E below. The air entirely surrounding it began to change into water, which dropped into the beaker from the pointed part.[23]

This somewhat crude instrument was refined and used by the Accademia throughout its short existence (1657–1667). Several variations were developed during the next two hundred years, but it was never a popular hygrometer. Charles Le Roy, a professor of medicine at Montpellier, however, apparently used it to formulate the concept of dew-point. In 1751, he described the results of his hygrometric experiments to the Royal Académie des Science in Paris.[24] His experimental device consisted of a tin vessel containing water in which a thermometer was immersed. The temperature of the water was gradually lowered by introducing ice. Le Roy noted that at a certain temperature, water vapor condensed upon the vessel's exterior. This enabled Le Roy to determine the amount of invisible moisture in the surrounding air: "There is at all times a certain degree of cold at which the air is ready to release part of the water that it holds in solution. I call this temperature the 'degré de saturation' of the air."[25] Le Roy's instrument contained too many defects to be useful as a meteorological tool but his work was the first attempt to determine the dew-point for hygrometrical purposes.[26]

The standard method of obtaining humidity measurements today is by using wet and dry bulb thermometers. Although not perfected until the nineteenth century, it was first developed in the middle part of the eighteenth century. In 1755, William Cullen, Professor of Medicine at Edinburgh and one of Joseph Black's teachers, published "An Essay on the Cold produced by Evaporating Fluids, and of some other means of producing Cold."[27] Working on a lead provided by one of his students, who had observed that when a thermometer was first immersed for some time in alcohol and then redrawn, the mercury fell two or three degrees, Cullen performed several experiments to confirm his conjecture that a wet thermometer cooled because of evaporation. He worked with several other liquids besides water and compiled a list of the comparable degree of cooling by the thermometer when it was immersed and then withdrawn from them. The power of a liquid to produce cold on evaporation appeared both to be proportional to its volatility and to depend on such factors as air agitation and warmth. Although Cullen did not suggest the use of the wet bulb as a hygrometer, he was the first to publish the correct explanation for the cooling of wet thermometers.[28]

In 1799, Sir John Leslie described an instrument which anticipated the wet and dry bulb hygrometer.[29] His instrument consisted essentially of a U-tube whose upper ends terminated in a hollow bulb. The tube contained slightly more than enough colored sulphuric acid to fill the whole of one leg, and one of the bulbs was covered with wet muslin. The cooling caused by evaporation from the muslin contracted the air with a consequent rise in the level of the liquid in the corresponding leg of the tube. As the displacement of equilibrium was greater the drier the air, the readings on the tube indicated not only the rate of evaporation, but also the humidity of the air.

Thus, by the beginning of the nineteenth century the stage was set for the development of the psychrometer as a truly scientific meteorological instrument. The first psychrometric observations were made at Carlsruhe by Prof. Boeckmann in the summer of 1802, using a Leslie hygrometer.[30] Several nineteenth-century scientists were prominent in the development and perfection of the psychrometer, including E.F. August and J.A. Mason.[31] G.J. Symons names seventy-two men who worked on the development of the psychrometer and other forms of the hygrometer between 1800 and 1880.[32] Their combined efforts made the hygrometer a valuable tool in the growth of meteorology as a science.

Two other instruments still currently used by meteorologists were also developed before 1800—the rain gauge and the anemometer. In theory the rain gauge is probably the simplest of meteorological instruments, yet a great deal of research and experiment has gone to make it an accurate instrument for measuring rainfall. For example, in 1769, Heberden demonstrated that a rain gauge on top of a 30-foot tall house caught only four-fifths as much rain as the gauge at ground level, and a gauge at the top of a 150-foot high abbey tower received only just over half the ground-level catch.[33] The cause of this loss at higher altitudes, which was due to eddies around the exposed gauges in strong winds, was not fully understood until the latter part of the nineteenth century.

The first known account of a rain gauge occurs in the manuscript *Arthastra* by the Indian Kautilya, ca. 400 B.C.[34] The rain gauge was simply a bowl with a diameter of about eighteen inches. Rainfall measurements were regularly taken by the Indians as an important aid in determining the annual crop to be sown. There is no indication, however, the Indians of this period thought of rainfall in terms of depth of water. This idea apparently originated around 100 A.D. in Palestine.[35]

Nearly 1,400 years later another rain gauge used in agriculture appeared in a fifteenth-century Korean manuscript:[36]

> In the 24th year (1442) of the reign of King Sejo, the King caused a bronze instrument to be constructed, in order to measure the rain. This is a vase (30 cm) in depth and (14 cm) in diameter, standing on a pillar. The instrument has been installed at the Observatory, and each time that rain falls, the officials of the Observatory measure the height with a scale, and make it known to the King. These instruments were distributed in the provinces and cantons, and the results of the observations were sent to the court.

The first use of a simple rain gauge in Europe was by

Benedetto Castelli.[37] In a letter to Galileo on June 18, 1639, Castelli spoke of having been in Perugia when it rained continuously for eight hours.[38] Desiring to estimate the amount the rainfall contributed to the outflow of Lake Trasimeno and assuming a uniform distribution of rainfall Castelli performed an experiment:[39]

> . . . , taking a glass vessel in the form of a cylinder, about a palm and half a palm wide, and putting in a little water, enough to cover the bottom of the vessel, I carefully noted the level of the water, and then exposed it in the open air to receive the rainwater that might fall into it, and let it remain for one hour.

He then indicated with a line segment in his letter how much the water level had risen in the vessel. This was certainly a case of the scientific spirit in action.

The earliest English instrument for measuring rainfall was designed by Sir Christopher Wren in 1662.[40] Its water container emptied when filled to a certain height. About fifteen years later Richard Townley invented a rain gauge which consisted of a funnel, 12 inches in diameter, soldered to a pipe which carried the water into a vessel where it could be weighed.[41] Townley also maintained the first extensive record of rainfall in England from 1677 to 1703.[42] In 1695, Robert Hooke designed a similar instrument.[43] These were the type used at Gresham College in London for several years.

The earliest approximation to the modern day rain gauges was made in 1722 by the Rev. Mr. Horsley, of Widdrington, Northumberland:[44]

> The weighing the water and reducing it from weight to depth seemed pretty troublesome, even when done in the easiest method: to remedy this inconvenience (besides a funnel and proper receptacle for the rain)[45] I use a cylindrical measure and gauge. The funnel is 30 inches in diameter, and the cylindrical measures exactly 3 inches, the depth of the measure is 10 inches, and the gauge of the same length with each inch divided into 10 equal parts; or, instead of a gauge, the inches and divisions may be marked on the side of the cylindrical measure. The apparatus is simple and plain, and it is easy to apprehend the design and reason of the contrivance; for the diameter of the cylindrical measure being just 1/10 of that of the funnel, and the measure exactly 10 inches deep, 'tis plain that 10 measures of rain make an inch in depth, one measure 0.1 inch, one

> inch on the gauge 0.01 inche, and 1/10 of an inch on the gauge 0.001 inch. By this means the depth of any particular quantity which falls, may be set down with ease and exactness, and the whole at the end of each month or year may be summed up without trouble.[46]

Little improvement on Horsley's instrument was made until the middle of the nineteenth century.

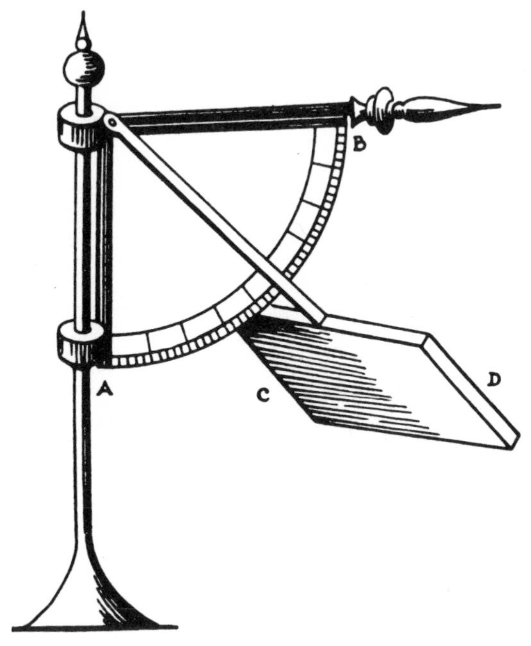

Hooke's Anemometer
Fig. 6.5

Although most of the work on the anemometer, which seeks to measure both the direction and the force of the wind, took place after 1800, pressure-plate and pressure-tube anemometers were developed earlier.[47] The pressure-plate anemometer, the first wind-measuring instrument, was first described around 1450 by the Italian mathematician, Leon Battista Alberti.[48] In his treatise "On the pleasures of mathematics,"[49] Alberti described a weathervane-like instrument with a little swinging plate on its tail. The purpose of the vane was to keep the small plate facing the wind. Behind the small plate was an arc with a scale to measure the deflections of the plate. Apparently Leonardo da Vinci described a similar instrument around 1500[50] and is often given credit for its invention.[51] The current evidence, however, points to Alberti as the inventor.[52]

In 1667 Robert Hooke constructed a practical instrument employing the swinging plate to measure the force of the wind (*Fig. 6.5*).[53] Partly because of its simplicity, Hooke's anemometer enjoyed some popularity for over a hundred years before it was replaced in the nineteenth century by the rotational and pressure-tube types of anemometers.

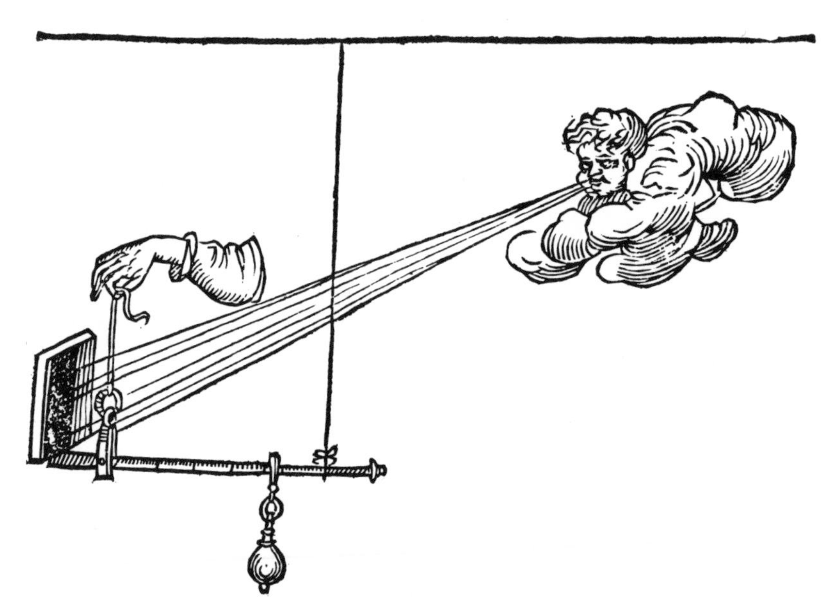

Santorio's Anemometer
Fig. 6.6

In 1625 Santorio Santorre, Professor of Medicine at Padua, had developed a different pressure-plate anemometer, which consisted of a flat plate attached to a scaled bar (*Fig. 6.6*).[54] The weight on the right would be moved, much as a standard scale, with changes in the wind's force. (There seems to be no reason for the string joining the bar to the top of the page.) In the middle of the eighteenth century, Pierre Bouguer employed Santorre's concept to construct a light, portable anemometer to measure winds at sea.[55] This instrument consisted of a 6-inch square piece of cardboard attached perpendicularly to a lighter rod, which pressed into a tube against a spring (*Fig. 6.7*). This tube was to be held in the hand with the face of the cardboard presented to the wind; the pressure of the wind was shown on a scale on the rod.

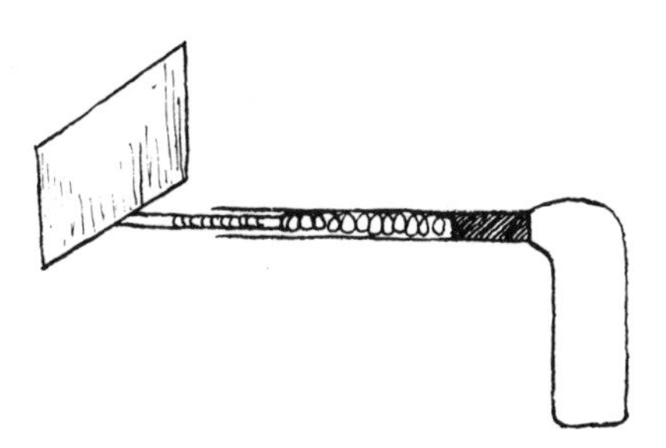

Bouguer's Anemometer
Fig. 6.7

The pressure-tube anemometer was another instrument designed to measure the "pressure" of the wind.[56] The earliest instrument of this type was described by a French courtier and ecclesiastic, Pierre Daniel Huet, sometime before 1721.[57] The first developer of an anemometer of this type to be considered a truly meteorological instrument, however, was James Lind, a physician from Edinburgh. In 1775, Lind described his "wind-gage."[58] It consisted of two glass tubes AD, CB, each about 6 inches in length, connected to each other by a bent glass tube AB. A short brass tube with its mouth F turned outwards was mounted on the tube AD. To allow the mouth F to be always pointed into the wind, the entire instrument was mounted on the spindle KL. The tubes were half filled with water to measure the wind's force, and the position of the scale was adjusted until the zero at its center coincided with the common level of the two water surfaces. The pressure from the wind would cause the water in tube AD to be depressed with a corresponding elevation of the water in tube CB. The sum of these displacements would indicate the force of the wind.

These early pressure-tube anemometers were plagued with accuracy problems. It was not until the end of the nineteenth century that W.H. Dines constructed a pressure-tube anemometer of sufficient accuracy to be used extensively in meteorology.[59]

Although the development of instruments to indicate wind direction, as distinct from wind force, dates back to ancient times,[60] the problem of measuring wind velocity still presents

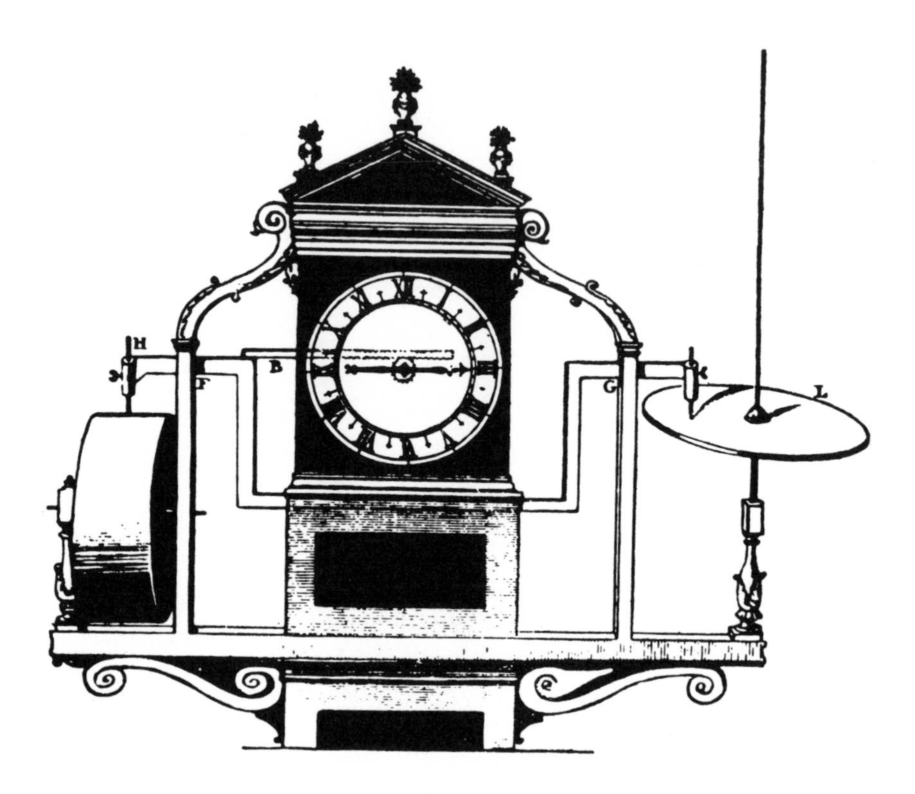

Weather-Clock
Fig. 6.8

problems because of the irregular nature of airflow. Thus, the anemometer can be considered both one of the oldest, and one of the newest instruments in the meteorologist's arsenal.

To conclude this chapter, mention should be made of an instrument developed in the seventeenth century which was a forerunner of the modern meteorographs—the weather clock (*Fig. 6.8*). There is some confusion over who should be given credit for the instrument shown in Fig. 6.16. Middleton holds that this instrument was designed as early as 1663 by Sir Christopher Wren.[61] It was Robert Hooke, however, who apparently first constructed a workable model of a weather clock between 1673 and 1678.[62]

The weather clock consisted of two parts: a strong pendulum clock which, in addition to showing the time, turned a paper-covered cyclinder and operated a mechanism for making punches on the roll of paper once every fifteen minutes, and a set of instruments consisting of a barometer, thermometer, hygroscope, windmill whose revolutions were counted, and a rain bucket. The changes in the readings of these instruments were recorded every quarter hour on the paper-covered cylinder by the above mentioned mechanism. This weather clock was so complicated that it spent much of its time laid up for repairs. Nevertheless, it illustrated the great vision seventeenth- and eighteenth-century meteorologists had of the future of meteorological instrumentation.

References

1. Aristotle, *Meteorologica,* Trans. H.D.P. Lee (Cambridge, Mass.: Harvard University Press, 1952), pp. 73–74.

2. *Ibid.,* p. 71.

3. René Descartes, "Les Météores," *Discours de la Methode* . . . (Paris: Charles Angot, 1668), pp. 227-230.

4. For a translation of the pertinent paragraphs of this work, see James R. Newman, ed., *The World of Mathematics,* 4 Vols., (New York: Simon and Schuster, 1956), 2:774–777.

5. W.E. Knowles Middleton. *Invention of the Meteorological Instruments* (Baltimore: The Johns Hopkins Press, 1969), p. 82. For a more thorough discussion of the developing theories concerned with water vapor. see W.E.K. Middleton, *A History of the Theories of Rain* (London: 1965), especially Chapters 1, 2, and 7.

6. Harvey A. Zinszer, "Meteorological Milepost," *Scientific Monthly* 58 (1944): 261.

7. Florian Cajori, *A History of Physics* (New York: The Macmillan Company, 1906), p. 48.

8. A picture of this instrument, along with a discussion of it, appeared in the work: Joachim d'Alencé, *Curieux traité de mathematique, ou par le moyen de trois instruments, a sauvoir, du barometre,* . . . (Paris: 1713), pp. 117–120.

9. Middleton, *Invention of the Meteorological Instruments,* p. 86.

10. There is some evidence to indicate that A. Mizaldus observed this property as early as 1554; see G.J. Symons, "A Contribution to the History of Hygrometers," *Quart. Jour. of the Roy. Meteor. Soc.* 7 (1881): 161.

11. Robert Hooke, *Micrographia* (London: Martyn and Allestry, 1665), pp. 147–152.

12. Middleton, *Invention,* pp. 86–88.

13. Friedrich Lowenhaupt, "Johann Heinrich Lambert als Naturforscher," *Johann Heinrich Lambert: Leistung und Leben* (Mulhausen: Braun & Co., 1943), p. 35.

14. G.J. Symons, "A Contribution to the History of Hygrometers," *Quart. Jour. of the Roy. Meteor. Soc.* 7 (1881): 162.

15. Johann H. Lambert, "Suit de L'Essai d'Hygrometrie," *Nouveaux Mémoires de L'Académie Royal Des Sciences* (Berlin, 1772): 65–102.

16. *Ibid.*, p. 75.

17. A. Wolf, *A History of Science, Technology, and Philosophy in the 18th Century* (New York: The Macmillan Co., 1939), p. 335.

18. Symons, *op. cit.*, p. 166. This type of hygrometer was first suggested by Guillaume Amontons in 1687. See W.E.K. Middleton, *Invention of the Meteorological Instruments*, pp. 48–99.

19. Wolf, *op. cit.*, p. 334.

20. Horace B. De Saussure, *Essais sur l'Hygrométrie* (Neuchatel: 1783).

21. Wolf, *op. cit.*, p. 327.

22. For a discussion of this debate, see Middleton, *Invention of the Meteorological Instruments*, pp. 103–110.

23. Middleton, *Invention*, p. 110.

24. Charles Le Roy, *Mém. Acad. Roy. des Sci. Paris* (1751), pp. 481–518.

25. *Ibid.*, p. 490.

26. Symons, *op. cit.* p. 166.

27. William Cullen, "An Essay on the Cold produced by Evaporating Fluids, and of some other means of producing Cold," *Edinburgh Philosophical and Literary Essays* II (1755): 159–171.

28. Middleton, *Invention*, p. 122.

29. Wolf, *op. cit.*, p. 341.

30. Symons, *op. cit.*, p. 170.

31. Middleton, *Invention*, pp. 126–128.

32. Symons, *op. cit.*, p. 181.

33. Geoffrey Reynolds, "A History of Rain Gauges," *Weather* 20 (April, 1965): 106.

34. For the related passage in this work *Arthastra*, see Josindra Nath Sammadar notes in *Quart. Jour. of the Roy. Meteor. Soc.* 38 (1912): 65–66.

35. Middleton, *Invention*, p. 134.

36. *Ibid.*, pp. 134–135.

37. Castelli is sometimes credited with the invention of the rain gauge; see L. Dufour, "Les grandes époques de l'histoire de la météorologie," *Ciel et Terre* 59 (1943): 358.

38. Galileo, *Opere*, ed. nuz., 20 vols., (Florence: 1890–1909), 18:62–66.

39. *Ibid.*, p. 62.

40. G.J. Symons, "A Contribution to the History of Rain Gauges," *Quart. Jour. of the Roy. Meteorol. Soc.* 18 (1891): 128.

41. Richard Townley, *Phil. Trans.* 18 (1694): 52.

42. Middleton, *Invention*, p. 137.

43. A. Wolf, *A History of Science, Technology, and Philosophy in the 16th and 17th Centuries* (London: Allen & Unwin, 1935), p. 310.

44. *Ibid.*

45. The common means of measuring the rainfall at this time was by weighing it, i.e., the Gresham College Rain Gauge.

46. See the short note by Horsley in *Phil. Trans.* 32 (1723): 328–329.

47. The rotational anemometer was a product of the nineteenth century; for a discussion of the development of the rain gauge and atmometer in the nineteenth century, see Middleton, *Invention*, pp. 141–174, 212–224.

48. Middleton, *Invention*, p. 182.

49. Leon Battista Alberti. *Opuscoli morali . . . tradotti . . . da cosimo Bartoli* (Venice, 1568), p. 253.

50. Ivor B. Hart, *The Mechanical Inventions of Leonardo da Vinci* (London: 1925), pp. 25–26.

51. J. Oliver, "An Early Self-Recording Pressure-Plate Anemometer," *Weather* 12 (1957): 16.

52. Middleton, *Invention*, pp. 182–183.

53. J.K. Laughton, "Historical Sketch of Anemometry and Anemometers," *Quart. Jour. of the Roy. Meteor. Soc.* 8 (1882): 162–164.

54. S. Santorre, *Sanctorii Sanctorii Iustinopolitani . . . commentaria in primam fen primi libri canonis Auicennae . . .* , (Venice: 1625), Col., pp. 245–246.

55. P. Bouguer, *Traité du navire, de sa construction, et de ses mouvemens* (Paris: 1746), pp. 359–360.

56. One reason for the great interest in the wind's pressure was that, especially for sailors, the pressure of the wind was considered its most important property.

57. Middleton, *Invention*, pp. 191-193.

58. James Lind, *Phil. Trans.* 65 (1775): 353-365.

59. See Middleton, *Invention*, pp. 196-201.

60. The "Tower of the Winds" contained a crude form of a wind vane on its top.

61. Middleton, *Invention*, pp. 248-250.

62. R.T. Gunther, *Early Science in Oxford*, (London: Dawsons of Pall Mall, 1967), 7:519-523.

Chapter Seven

Meteorological Observations

The development of the instruments discussed in the preceding chapters was necessary to overcome the impediment which had developed in the fifteenth century to the evolution of scientific meteorology. The advent of the thermometer, barometer, hygrometer, etc., as scientific instruments, opened the way for a more comprehensive study of the atmosphere.

The first step was, of course, the gathering of scientific data, which involved the taking of meteorological observations. This was nothing new. There is little doubt that weather observations were made and recorded throughout antiquity. Certainly Theophrastus would not have been able to develop his empirical rules for weather prognostication without them. The Greeks, in fact, were one of the first to make regular meteorological observations from as early as the fifth century B.C.[1] The results were exhibited to the public on the so-called parapegna, a form of almanac fixed on columns. Observations on wind predominated, doubtless because of the importance of wind information for maritime travel.[2]

Ancient weather observations were naturally quite crude compared to modern. They consisted mainly of sky conditions (cloudy, rain, clear, etc.), wind direction, the "warmth" of the air (warm, hot, cold, etc.), and sometimes the amount of precipitation, apparently the first of the major elements to become systematically recorded.

The taking of meteorological observations in this early period never completely ceased, despite long and frequent interruptions. It was the custom of Roman historians to note in their annals the more important atmospheric phenomena, especially those necessitating sacrifices,[3] a custom handed down to the chroniclers of the Middle Ages. The quality of these weather entries so improved that by the end of the thirteenth century the character of the seasons during these periods could be ascertained. According to Hellmann, the fourteenth-century Englishman William Merle, a fellow of Merton College, was the first man in the occidental world to keep a regular day-by-day weather journal. These observations were taken at Oxford from 1337 to 1344.[4] Another example of observations made in the fourteenth and fifteenth centuries was found in a manuscript in the university library at Basel[5] for the years 1399 to 1406. The entry for September 5, 1399, was: "A.M., fair and clear; P.M., cloudy and hot; after sunset dark clouds with coruscations and a little thunder, in the night heavy rain."[6]

The sixteenth and early seventeenth centuries saw a continued increase in the attention given by European scientists to meteorological observations. The mathematician and astronomer Johann Werner (1468–1528) was the first German to take regular meteorological observations, doing so from 1513 to 1520.[7] Others who devoted their efforts to taking regular weather observations included the Scandinavian astronomer Tycho Brahe (1546–1601), whose observations from 1582–1598 comprised one of the earliest systematic meteorological records,[8] and the renowned German astronomer and mathematician, Johann Kepler (1571–1630). Around 1600, Kepler started a weather diary which he kept up for most of his life. He also made regular meteorological observations in the city of Prague from 1604, and in Sagan from 1628.[9] None of the meteorological observations hitherto contained temperature, pressure nor humidity observations, however, and it was not until well into the seventeenth century that these important observations first appeared in weather records.

Many leading seventeenth-century scientists took observations with the new meteorological instruments. Galileo and his friends attempted to take temperature readings as early as 1613.[10] René Descartes, who despite his outwardly obstinate assertion that his concept of the composition and operation of the atmosphere was valid, realized the need for more atmospheric knowledge, and gave his complete support to the increasing attempts to employ these instruments in the study of the atmosphere.[11] The earliest recorded meteorological observations with instruments were made in concert at Paris, Clermont-Ferrand, and Stockholm, between 1649 and 1651 to test the capacity of the barometer to forecast the weather.[12] In Stockholm, they were first taken by Descartes himself.[13] Among the other participants were Boyle, Hooke, Mercenne, and Leibniz, at whose instigation observations of barometric pressure and weather conditions were also made at Hanover in 1678, and at Kiel from 1679 to 1714.[14]

In spite of the great names involved, these attempts with the new meteorological instruments were scattered and fragmentary.[15] Efforts to form some sort of international weather-observing network did not make much headway until the eighteenth century. The forerunner of these networks, however, occurred in the middle of the seventeenth century in Italy when Ferdinand II of Tuscany made meteorological observation one of the activities of the Accademia del Cimento. He had instruments, mainly thermometers, barometers, and hygrometers, constructed and sent to chosen observers in Florence, Pisa, Vallombrosa, Curtigliano, Bologna, Milan, and Parma. Later the network was enlarged to include the cities of Paris, Osnabruck, Innsbruck, and Warsaw, all of which gave it an international character. Observations of temperature, pressure, humidity, wind direction, and state of the sky were entered on forms, and subsequently sent to the academy for comparisons with each other. Unfortunately, due to ecclesiastical pressure, the academy was forced to dissolve after only ten years of activity on July 14, 1667. Nevertheless, the Accademia del Cimento had an important influence on the development of weather-observing networks of the eighteenth and nineteenth centuries.[16]

Up to the middle of the seventeenth century there had been no standard method for taking meteorological observations. The Accademia del Cimento during its short life had made an attempt at such a standardization and Robert Hooke took another step in this direction. Realizing the importance of standardization for any further study of the atmosphere, Hooke, in 1663, proposed a "Method for Making a History of the Weather":[17]

> For the making a more accurate history of the changes of weather, it will be requisite to observe:
> 1. The severall winds, what quarters they blow in, how long in each quarter, and with what strength at several times of the Day or in the severall parts of the time of their duration; as whether mild at first when the wind begins to blow, in such a gravity stronger about the midst of its duration, and slow again at the changing, as we find in the current or tydes of ye sea; or whether the wind, when blowing very stiff, does suddenly change into another quarter and blow as stiff as was observed in the last great wind; and this is to be observed with a weather cock placed in some very high place with particular contrivances that may give a certain estimate of the quarter and of the strength of ye wind.

> 2. The Moisture and dryness of the air with the degrees of it. And this to be observed with a good hygroscope which may be made either with the beard of wild Oat, a gut string or the like. But I prefer the beard of an oat because that, though often wet, if dryd againe it will twist itself as much as before, which will not happen if experiment be made with a gut string. I want further to have it made with a single one because there may be the better comparison made, as I shall shew anon.
> 3. The heat and cold of ye air with its degrees and continuances, which may be observ'd by a good thermometer seald up soe that it may serve for a constant standard both for Winter and Summer from year to year, as long as the Observations be made.
> 4. The Greater or lesser pressure of ye air, that is, the differing height of ye mercuriall Cylinder; and this to be observd with an instrument that may shew the last variation of that kind, which may be contrived severall ways.
> 5. The Constitution and face of the Sky, or heaven; that is, whether clear and blew or thick; if thick, after what manner, whether with a thin whiteness, cloudy, etc.; what raines, mists, foggs, snows, hailes; which way the clouds drive—whether an other way, or the same way the wind passes. The colour and face of the sky and clouds. . . .

An example of the format proposed by Hooke in recording weather observations is given in *Fig. 7.1.* It is interesting to observe that Hooke did not advocate taking observations at any set time during the day, but only at those hours in which a significant change in weather occurred; as long as at least one observation was made each day.

This paper was presented to the Royal Society and copies sent to several persons engaged in the taking of meteorological observations. It marked the first attempt to precisely describe what should be included in a weather observation, and how, using *standard* instruments, the observations were to be made. Hooke's proposal is considered by several historians the turning point in meteorological development, which provoked a reevaluation in meteorological thought, culminating in the air-mass and frontal analysis techniques developed by the V. Bjerknes school in the early 1900's.[18] For many, Robert Hooke is the Father of Modern Meteorology.

The scattered attempts to compare meteorological observations taken simultaneously at a number of different places in-

The *Form* of a *Scheme*.

Which at one view represents to the Eye Observations of the Weather, for a whole Month, may be such, as follows.

Days of the Moneth, and Place of the Sun	*Remarkable hours.*	*Age and Sign of the Moon at Noon.*	*The Quarters of the Wind, and its strength.*	*The Faces or visible appearances of the Sky.*	*The Notablest Effects*	*General Deductions.* These are to be made after the side is filled with Observations, as
June 14 ♊ 12.46'	4	27	W - - - 2	Clearblue, but yellowish in the *N E.*	A great Dew	From the last Quarter of the Moon to the Change, the weather was very temperate, but for the Season, cold; the Wind pretty constant between *N.* and *W.* &c.
	8		- - - 3	Clouded toward the *South.*		
	12	♉ 9. 46	- - - 3½		Thunder far to the S.	
	4		- - -	Checkered blue.		
	8	*Perigeum*	WSW 1		A very great Tyde.	
	12		- - -			
15 ♊ 13.40'	8	28	N W 3	A clear sky all day, but a little checker'd about 4 *P. M.* At Sun-set red and hazy.	Not by much so big a Tyde as yesterday. A great Thunder-Showre from the *N.*	
	4		4			
	6	♊ 24. 51	N 2			
	12		1			
16 ♊ 14. 57 &c.	10	*New Moon at* 7. 25. *A. M.* ♊ 10.8 &c.	S 1 &c.	Overcast and very lowring, &c.	No dew upon the ground, but very much upon Marble-stones, &c.	

Hooke's Weather Observation Scheme

Fig. 7.1

creased during the eighteenth century. Their value was greatly enhanced by improvements in the design, construction, and standardization of meteorological instruments, and the growing insistence upon following uniform procedures in making the observations. The scientific societies of this century played an increasingly important role in these efforts.

The French Academy of Sciences, founded in 1666, soon began to make regular meteorological observations of the Academy's observatory in Paris. These efforts were systematized, and by 1688 the Academy had a record constantly kept by one of its members. Summaries, which included pressure, temperature, and rainfall readings, were published yearly in the Academy's memoirs. One of the members who maintained meteorological records was the mathematician Philippe De La Hire (1640–1718).[19]

In this summary for 1709, De La Hire began by listing the monthly amount of rainfall or melted snow at the observatory:

> The quantity of water which fell, either in rain or melted snow was

	Lin.		Lin.
Jan.	22 3/8	July	18 2/8
Feb.	13 7/8	Aug.	10 7/8
March	20 6/8	Sept.	29 2/8
April	37 6/8	Oct.	17 5/8
May	32	Nov.	1 3/8
June	45 4/8	Dec.	11 6/8

> The sum of the water of the whole year 1709, is 261 lines 1/8, or 21 inches, 9 lines 1/8, which is a little more than the mean years, which we have determined to be 19 inches.

Next, he discussed in general terms the weather patterns for the year 1709:

> The cold, at the beginning of this year, was excessive, with a great deal of snow; for my thermometer fell to 5 parts, the 13th and 14th of January; and the following days being a little risen, it returned to 6 parts the 20th, and the 21st to 5¼, but afterwards the cold diminished gradually.[20]

Note that there were no systematic recordings of the temperature or atmospheric pressure—probably the result of a lack of agreed standards for thermometric and barometric scales (see Chapters IV and V). The rain gauge was a tin vessel four square feet in area and six inches deep, with a bottom that sloped slightly to one corner where a short tube led the rainwater into a jug. After each rain the water was measured in a small cubical vessel.

In 1717 a German doctor, Johann Kanold, attempted to form

another international weather-observing network. Kanold induced a number of observers in Germany as well as several abroad to send him their recorded observations. He compiled and published them in a quarterly journal known as *Breslauer Sammlung,* beginning in 1717.[21] This arrangement lasted only for ten years, however, when development of international meteorological observation shifted from the Continent to England.

In 1723, following the lead of his predecessor, Robert Hooke, James Jurin, Secretary of the Royal Society, invited all weather observers who were both willing and equipped for the work, to submit annually to the Society the records of their daily meteorological observations. Included in his request were detailed instructions on how these observations were to be made and recorded. According to the procedure outlined by Jurin, the observers were to record at least once a day the readings of their barometer and thermometer, the direction and force of the wind, the quantity of rain or snow-water collected since the last observations, and the appearance of the sky. Special barometric observations were to be made during any severe storm, to consist of the time of the storm and barometric readings at its start, height, abatement and end. The observations were to be recorded in journals having six parallel columns, entries in which should show the date and hour of observation, the barometric reading, the temperature, the direction and force of the wind (the force of the wind was to be estimated according to a scale of four degrees ranging from 0 representing calm conditions to 4 representing the most violent wind), a concise description of the weather, and amount of rain or snow-water (in inches and tenths of an inch) collected since the last observation. The averages of the barometric reading and the temperature, along with the total amount of rain or snow-water collected since the last observation, were to be determined for each month, and then for the whole year. Each year the observers were to send copies of their journals to the Secretary of the Royal Society for comparison with each other, and with the Society's own weather journal. The results of this collection were to be published yearly in the *Philosophical Transactions.*[22]

In his proposal Jurin strongly stressed the importance of employing standardized instruments and recommended the type of barometer and thermometer to use. In the case of the thermometer, it was suggested that the thermometer should be located in a north room where a fire was seldom if ever lighted. The rain gauge should consist of a two-to-three-foot-in-diameter funnel, which emptied itself through a long stem into a graduated cylindrical measuring-vessel, kept as airtight as possible to reduce loss by evaporation. This instrument, of course, was to be situated in a completely unsheltered place.

The high reputation enjoyed by the Royal Society throughout the scientific world caused Jurin's appeal to receive widespread initial notice and resulted in the first truly international effort at collecting and comparing weather observations. For a few years, beginning in 1724, many weather observers sent copies of their journals to London. These observations came from such places as Upsal in Sweden, Abo in Finland, Naples, Rome, India, and North America. Among those who answered Jurin's appeal was the American mathematician Isaac Greenwood (1702–1745) (*Fig. 7.2*).[23]

By talking to various ship captains, in particular the captain and crew of a ship he once took from England to America, Greenwood learned that many ships maintained "Journals of Voyages." In these journals, he found three entries which he felt would be of meteorological value. First there was a general account of the weather for every day. Next, there were entries giving the direction, and often the force, of the winds, on observations taken and recorded every two hours. Thirdly, there was a daily account of the latitude and longitude of the ship.

Here Greenwood saw an opportunity to enlarge on Jurin's proposal for a worldwide land network of meteorological observation stations. Thus, in 1728, Greenwood proposed that the Royal Society collect as many present and past marine weather observations as possible and compile this information in table form for the different oceans.[24] From this he envisioned many advantages, both for meteorology and marine navigation. Such an undertaking could define the bounds and limits of all "considerable" winds, as well as the origin and path of movement of these winds. It might also be possible to determine the relationship, if any, between winds and marine weather. Another advantage would be a knowledge of what type of winds prevail at the various latitudes, and during the different months of the year, including the peak seasons of hurricane activity, as well as their various paths of movement, important information for marine navigation and the east coast of America.[25]

Honoured Sir, Boston N England. May 1. 1727

The Occasion of this Letter to You from One so little known by You, & of so obscure a Name is ye Indisposition of ye Reverend Dr Cotton Mather who would otherwise have wrote Himself.

The Doctor has been for a long Time endeavouring to engage his Friends in making Meteorological Observations that Our Country might not be ye most backward in following Your ingenious *Invitatio*, but ye unskillfulness of Some, & ye Business of others have I think hitherto prevented his Design, excepting only that He has so far prevailed upon an Ingenious Tradesman of this Town Mr Feveryear by Name, by presenting him with a Barometer, & some other things as to obtain ye Inclosed Observations.

My Charge from ye Doctor at present is, to acquaint You that ye Character of ye Author of these Observations is such that You may depend upon ye accuracy of them, and tho' there are Some few things admitted as for instance ye Degree or Strength of ye Wind what is inserted especially ye Barometrical Observations are performed with a great deal of Exactness; and I think have One thing peculiar to Them wch is that They are ye first Set of Such Sort of Observations that was ever made in New England.

I should have examined ye Tube, and Mercury with which these Observations were made but my Opportunity is so Short at present that I have not Time.

I shall add no more than just to observe to you that I sent last Fall to ye Rev. Mr Derham a large Collection of Observations on ye Weather, being those of Mr Robie's wch were sent to Me during my Residence at London, but having departed thence before They arrived, they were again returned to Me in N England, these were Observations for 10, or 12 Years successively wch may perhaps be of Some Service to your Design.

Also, That if it would be acceptable to ye R. Society to have an Annual Account of ye State of ye Weather in these parts of ye World, especially at Cambridge, that I shall have all ye Opportunities imaginable thereunto, being chosen by ye College their Hollisian Professor of ye Mathematicks & Experimental Philosophy wch place has some peculiar Advantages for Observation, above most of ye same Nature in ye world, being accommodated wth a very large Apparatus of Glasses, and other Instruments, besides by its Institution, furnish't wth 10 Pensionary Scholars of ye 2 upper Classes who will always be ready to continue on ye Observations in case of Sickness, absence, or any other Accident. If Likewise it may be of any Service to send out such astronomical Observations as we shall make, you may expect a constant account of such Eclipses &c. as may occur, being furnished with very good Instruments for that Design, Virt. ye same Quadrant that Dr. Halley had to observe on ye Southern Constellations at St Hallena [sic] besides several good Telescopes.

I am,

Sir,

Your most Obedient Humble Servt.

Isaac Greenwood

Dr. Jurin
Dated May 1 1727 To Dr. Jurin

Letter by Isaac Greenwood

Fig. 7.2

About fourteen years after Greenwood's proposal, Roger Pickering submitted to the Society a similar proposal, "Scheme of a Diary of the Weather, together with draughts and descriptions of Machines subservient thereunto":[26]

> A sense of the importance of observing the weather induced Hippocrates in his remarks upon the epidemic diseases in Thasos, to premise a general history of the weather preceding them; and with the same view did our great Mr. Boyle turn his thoughts so closely upon the same subject: whose example, being followed by several judicious inquiries into Nature, both abroad and at home, has brought the Natural History of the Air to a surprising degree of perfection, beyond what the Ancients ever could pretend to, or even thought of. Had but each county in England gentlemen of such sentiments who would charge themselves with the annual trouble of sending a regular account of the weather to this learned body, by it to be compared and digested, to what degrees of accuracy may we not suppose a knowledge of the nature and affections of the atmosphere may be brought; and how well may we not hope to be guarded against the disorders which as islanders we are exposed to, by such a close inquiry into the nature of that necessary fluid in which we breathe! not to mention the advantages which several important branches of trade may receive from such measures; and were the digested observations of the R. Soc. compared with those of foreign societies, formed upon the same plan, how short a time would bring this part of Philosophy to the greatest degree of demonstrable certainty.
>
> The trouble of making and keeping such meteorological registers, which, in all probability, prevents several gentlemen from performing this piece of service to the public, might be rendered very inconsiderable, by the proposal of an easy, as well as comprehensive, method for a diary, and a set of simple and convenient machines for making the necessary observations.

The instruments recommended by Pickering are shown in *Fig. 7.3*.

As was the case with those of Jurin and Greenwood, Pickering's proposal met with some initial enthusiasm but was then largely forgotten. The trouble was not that the societies were uninterested in such undertakings, but that there were so many other scientific areas calling for attention.[27] What was clearly needed was a separate society dedicated solely to meteorology.

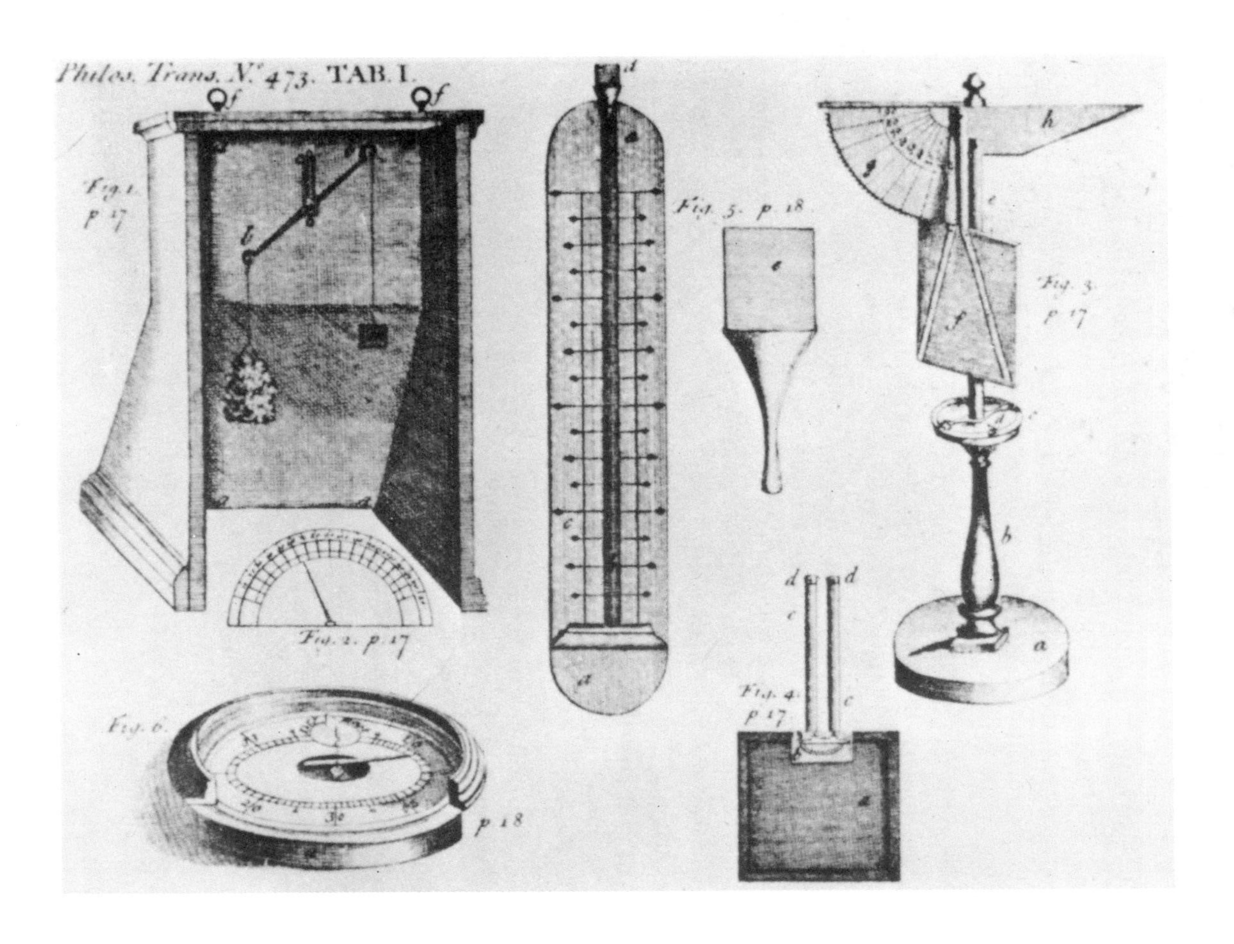

Pickering's
Meteorological
Instruments
Fig. 7.3

Although this was to be largely a product of the nineteenth century, its forerunner began in the late 1700s with the formation of the Society of the Palatina.

The Societas Meteorologica Palatine was founded in the German city of Mannheim in 1780 by Elector Karl Theodor of Bavaria[28] with headquarters in his castle at Mannheim. Its first director was the distinguished German meteorologist, J.J. Hemmer. One of the Society's first acts was to write to all the principal universities, colleges, and scientific academies soliciting their cooperation and offering to present them free with all the necessary instruments properly standardized. From the replies to this solicitation, which included thirty societies, fifty-seven institutions were chosen as observation stations, and along with the free instruments, detailed instructions and recording forms were sent to them. The instruments included a barometer, thermometers, a quill-hygrometer, a rain gauge, an electrometer, a windvane, and for some stations a magnetic needle. The wind force and the amount of cloud cover were to be estimated numerically on conventional scales. Observations were to be made three times a day, at 7 A.M., 2 and 9 P.M. This was truly an international organization. The fifty-seven observation sites extended from the Mediterranean, to North America, to Russia and Siberia (interestingly England and Siberia did not choose to participate). The collected information was compiled in the Society's *Ephemerides* and has been of considerable value in later climatological researches. The death of Hemmer in 1790, however, together with the political confusion in Europe following the French Revolution combined to cause the gradual collapse of the Society; its last volume (for 1792) appeared in 1795. It was not until well into the nineteenth century that a comparable organization arose to take its place.

From the earliest weather observations it was customary for the observer to use some sort of abbreviations for frequently recurring words, which were slowly replaced by symbols to represent common weather conditions such as rain, snow, thunder, etc. Each observer employed his own system, and there was no standardization. An example of symbols employed in the eighteenth century were those used by the meteorologist, Petrus Von Musschenbroek (1692–1761), in his printed record of weather observations made at Utrecht in 1728:[29]

ll	++	. .		-----
ll	++	. .		----
ll	++	. .		-----
(rain)	(snow)	(hail)	lightning)	

Elaborate standard systems were proposed near the end of the eighteenth century, but none were ever adopted.[30]

In 1771, J.H. Lambert proposed the worldwide taking of meteorological observations. To determine the observation sites, he pictured the world as an icosahedron, and placed one site in the center of each triangle. Additional sites were added in Europe and wherever Lambert thought the need warranted it. *Figure 7.4* shows Lambert's world map, the dots being the observation sites.

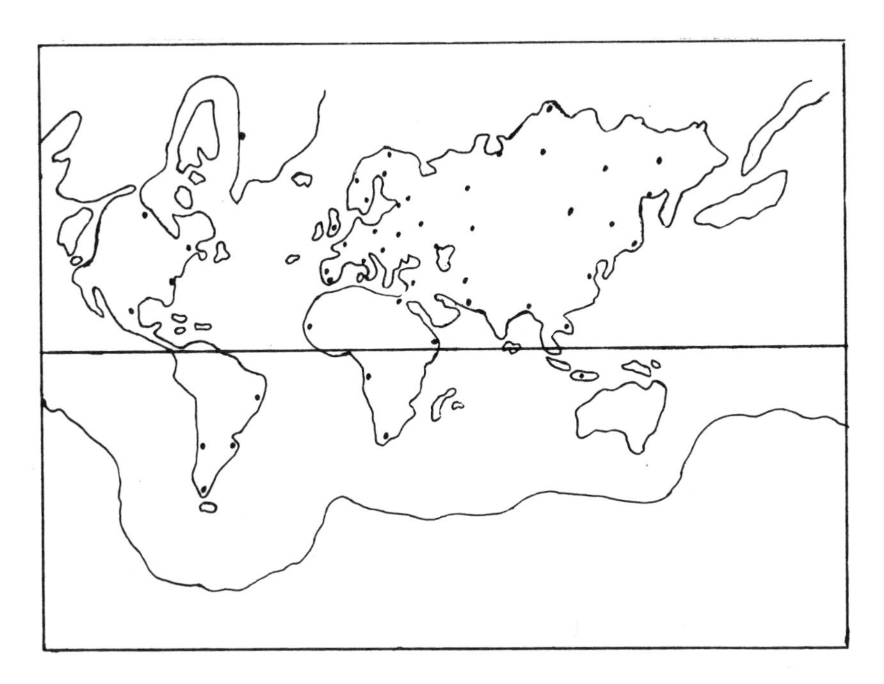

Lambert's World-Wide Meteorological Observation Proposal
Fig. 7.4

One observation a day was to be taken at noon London time to obtain a simultaneous, worldwide weather picture. Following the lead of Musschenbroek, Lambert proposed symbols to represent meteorological phenomena, such as (++) for snow. He noted that one often finds records of wind observations in which only the frequencies from different directions are given. Lambert proposed a method of finding the resultant direction of the wind which involved computing the vector mean from the direction frequencies by assuming a constant wind force for each direction. Despite its obvious limitations, this method was used for many years.[31]

By the end of the eighteenth century, the need for national and international meteorological observation networks had become clearly established. Initial efforts had been made by Hooke, Jurin, and Lambert, and organizations like the Accademia del Cimento and the Societas Meteorologica Palatine. The need for scientific societies and organizations dedicated solely to meteorology was recognized by scientific leaders, although such societies did not actually come into being until the middle of the nineteenth century. In addition, the ponderous amount of meteorological data compiled, compared, and studied during the seventeenth and eighteenth centuries provided a valuable foundation for meteorological research, whose fruits have so greatly matured during the past two hundred years.

For example, barometric observation led to the development of one of the more interesting problems in the history of meteorology—the pressure-height problem. The goal was to determine the relationship between atmospheric pressure and altitude and to apply this relationship to the measurement of mountain heights. The pressure-height problem became the most popular meteorological problem of the seventeenth and eighteenth centuries, and provided a strong catalyst in kindling scientific interest in the study of the atmosphere.

The story of the pressure-height problem began with the construction by Torricelli of the first barometer in 1643. The news of his experiment spread rapidly throughout Europe, and among the many scholars who received the news was M. Petit, chief of the French government's Department of Fortifications, who in turn passed the information to Blaise Pascal (1623-1662)[32], who became very interested and several times repeated Torricelli's barometric experiment, using different liquids, tubes, etc. The results of these experiments were published in a little pamphlet entitled "Experiences nouvelles touchant le vide," which was circulated throughout the continent and became famous in the European scientific world.[33]

Although the supposition that the pressure or weight of the

Blaise Pascal
(1623–1662)

atmosphere decreases with increased height is generally first attributed to Pascal,[34] there is some doubt whether he was, in fact, the first to discover this relationship. Some claim that Descartes suggested it to him, and others maintain that the idea came from Mersenne.[35] Torricelli is known to have suspected this pressure-height relationship, for he states in his first letter to Ricci:[36]

> We live submerged at the bottom of an ocean of the element air, which by unquestioned experiments is known to have weight,[37] and so much, indeed, that near the surface of the earth where it is most dense, it weighs (volume for volume) about the four-hundredth part of the weight of water . . . whereas that on the tops of high mountains begins to be distinctly rare and of much less weight than the four-hundredth part of the weight of water.

It was Pascal, however, who first proved that amospheric pressure decreased with an increase in height above the earth's surface. On September 19, 1648, he induced his brother-in-law, Florin Perier, to take readings of two barometers, at the foot and the top of the mountain Puy de Dome under the close observation of a select group of "distinguished ecclesiastical gentlemen and laymen."[38] Later that year, Pascal published the results in his *Recit de la grande experience de l'equilibre des liqueurs,* in which he suggested the use of the barometer in determining the height of different localities.[39] This suggestion resulted in a great deal of effort by many of the leading mathematicians of the following hundred and fifty years to establish a relationship between atmospheric pressure and height.

Pascal also gave an interesting mathematical conclusion for the total weight of the world's atmosphere—8,283,889,440,000,000,000 pounds. However, his discussion will not be presented here, for according to Pascal, "a child who knew how to add and subtract could do it."[40]

Thanks to men like Marin Mersenne, the work of Torricelli and Pascal became widely known in European scientific circles and provided the motivation for attempts to establish pressure-height relationships. Before any progress on this problem could occur, however, it was necessary to obtain some knowledge of the elasticity or compressibility of the air. This was provided in 1660 by Robert Boyle, who formulated what is now known as Boyle's Law: at constant temperature, the volume of a gas diminishes in the same ratio as the pressure upon it is increased. Boyle's formulation of this law, based on his experiments concerning atmospheric pressure,[41] provided the final inducement for the increased study of the pressure-height problem in the next 150 years.

Already deeply interested in nearly every phase of seventeenth-century meteorology, Hooke's attention was naturally drawn to the pressure-height problem.[42] His discussion of how atmospheric pressure varied with altitude appeared in his *Micrographia* (1665):[43]

> 1. Consider a vertical column of the atmosphere and divide it into one thousand layers all containing equal quantities of air.
>
> 2. By calculations from the density of the air at ground level, it is determined that each of these layers must contain as much air as is a column of air at the surface and 35 feet high.
>
> 3. Using Boyle's law, the heights of the layers can be determined, and are as follows:

Layer	Hgt. at top of Layer in Ft.
1	$35\frac{35}{999}$
2	$35\frac{35}{999} + 35\frac{70}{998}$
3	$35\frac{35}{999} + 35\frac{70}{998} + 35\frac{105}{997}$
. . .	. . . .
n	$35\frac{35}{999} + \ldots + 35\frac{n \times 35}{1000 - n}$

Hooke did not attempt to sum these thicknesses, and he recognized that the 1,000th layer must be of infinite thickness: "since we cannot yet find the 'plus ultra,' beyond which the air will not expand itself, we cannot determine the height of the air."[44]

Probably the most important mathematical concept which evolved from the attempts to investigate the variations of atmospheric pressure with altitude was that as the height increased in a geometrical progression, the pressure would correspondingly decrease in an arithmetical progression. The first person to realize this was Edme Mariotte (ca. 1620–1684).[45] In his "Discours de la nature de l'air,"[46] Mariotte divided the height of the atmosphere into 4,032 parts corresponding to as many layers of equal weights, each represented by one-twelfth of a line of the normal barometric height of twenty-eight inches. From experiments of his own, Mariotte concluded that the thickness of the lowest of these layers was five feet. He thus argued that the thickness of the 2,016th layer above the earth (under half the pressure at the surface) would be ten feet. Mariotte knew that the intervening layers would increase in geometrical progression, but, apparently wishing to avoid the difficulty of such a calculation, he adopted an approximating method in which he assumed that the average thickness of the intervening layers between the lowest and the 2,016th layer was the arithmetic mean of five feet and ten feet, i.e., seven and a half feet. This gave the height of the lower half of the atmosphere as 15,120 feet ($7\frac{1}{2} \times 2{,}016$). Similarly, Mariotte calculated the thickness of half the upper half of the atmosphere as also 15,120 feet. Although this process could be carried on indefinitely, Mariotte stopped after twelve successive applications of this approximating method.[47]

Mariotte was also the first to explain how the altitude of a high place, such as a mountain, could be computed with the barometer.[48] He did not develop a specific formula, but based his procedure on the assumption that a rise of sixty-three Paris feet resulted in the drop in the barometric reading of one line. Placing a barometer at each of two differently elevated locations, the altitude of the lower being known and the higher having to be determined, Mariotte employed the following formula: $H = 63d + 3/8(d - 1)/2$, where d was the difference in the readings of the two barometers. The $3/8(d - 1)/2$ was a correction factor that Mariotte obtained from examination of the data of the Puy de Dome experiment and a similar one performed by D. Cassini at a mountain in Provence.[49]

While Mariotte was the first to propose the theory that the atmospheric pressure decreased arithmetically as the altitude increased geometrically, the first to apply this concept to the problem of the relationship between height and pressure was the celebrated English mathematician and astronomer, Edmund Halley (1656–1742).[50] In his application of this concept, Halley made use of a relatively new mathematical tool, the logarithm.[51] The basis of his method was the analogy between Boyle's Law connecting the volume, or expansion, with the pressure of a quantity of gas, and the rule connecting the coordinates of a point on a hyperbola in relation to its asymptotes. Following this procedure, Halley formulated that

$$H = \frac{900\,(\log 30 - \log h)}{0.0144765}$$

where h is the mercury column height at the level in question; the height of the mercury column at sea level = 0.0144765, and 900 ft. the height of a cylinder of air equal to an inch of mercury at sea level.[52] Using this formula, Halley derived two pressure-height tables (*Fig. 7.5*).

The altitudes to given heights of the mercury		The heights of the mercury at given altitudes	
Inches	*Feet*	*Feet*	*Inches*
30	0	0	30.00
29	915	1000	28.91
28	1862	2000	27.86
27	2844	3000	26.85
26	3863	4000	25.87
25	4922	5000	24.93
20	10947	1 mile	24.67
15	18715	2 miles	20.29
10	29662	3 miles	16.68
5	48378	4 miles	13.72
1	91831	5 miles	11.28
0.5	110547	10 miles	4.24
0.25	129262	15 miles	1.60
0.1	154000	20 miles	0.95
0.01	216169	25 miles	0.23
0.0001	278338	30 miles	0.08
		40 miles	0.012

Fig. 7.5 Halley's Pressure-Height Relationship Table

He acknowledged that his formula was not precisely accurate: He had considered the "air and atmosphere as one unaltered body, as having constantly at the earth's surface the 800th part of the weight of water, and being capable of rarefaction and condensation in infinitum.[53] Halley realized that such factors as temperature variations influenced the density of air,[54] but he felt that his formula would suffice for the altitudes which would generally be under consideration.

By the beginning of the eighteenth century a great deal of scientific interest had been focused on how to use the barometer to determine the altitude of locales. The pressure-height formula of Halley's was an oversimplification of final formulae developed in the nineteenth century. However, it provided the foundation for the further exploration of the pressure-height problem by such scientists as Johann Lambert, Jean Andre De Luc, Johann Hennert, and Pierre Laplace.[55]

It should be noted that the correct solution to the pressure-height problem is obtained by combining the hydrostatic equation ($\partial p/\partial z = -g\rho$) and the ideal gas law ($p = \rho RT$) to obtain the equation

$$\frac{1}{p}\frac{\partial p}{\partial z} = -\frac{g}{RT}$$

which demonstrates dramatically that the rate of decrease of pressure depends on temperature. The modern solution (the hypsometric equation) is obtained from this relation by reversing the roles of p and z as dependent and independent variables to give

$$\frac{\partial z}{\partial p} = -\frac{RT}{g}\frac{1}{p}$$

Integrating gives

$$h(p_2) - h(p_1) = -\int_{p_1}^{p_2} \frac{RT}{g}\, d\ln p = \frac{R\overline{T}}{g} \ln\left(\frac{p_1}{p_2}\right),$$

where $\overline{T}$ is a suitable average temperature. Incidentally, the pressure-height relation survives not as a way to measure heights of mountains, but in two other contexts—to measure the altitude of airplanes (the aneroid barometer) and as an important component of modern synoptic evaluation. Directly, the mean temperature in a layer between two isobaric surfaces controls the thickness; indirectly, if thickness can be predicted, then the mean temperature is known and rain-snow decisions can be made.

References

1. Sir William Napier Shaw, *Manual of Meteorology* (Cambridge: The University Press, 1926), 1:11.

2. There is no evidence that the Greek Tower of the Winds, mentioned in Chapter Three, was used as an observation station. Rather this temple was probably a place where the devout could offer prayers and gifts in view of obtaining the wind and weather most desired for agricultural and nautical purposes. See Richard Inwards, "Meteorological Observations," *Quart. Jour. of the Roy. Meteor. Soc.* 22, No. 98 (1896): 81-84.

3. Gustav Hellmann, "The Dawn of Meteorology," *Quart. Jour. of the Roy. Meteor. Soc.* 34 (1908): 230.

4. *Ibid.*, p. 231.

5. Lynn Thorndike, "A Weather Record for 1399-1406 A.D.," *Isis* 32 (1940): 304-323.

6. *Ibid.*, p. 306.

7. Gustav Hellmann, *Die Anfänge der Meteorologischen Beobachtungen und Instrumente* (Berlin: 1890), p. 5.

8. Harvey A. Zinszer, "Meteorological Mileposts," *Scientific Monthly* 17 (1944): 262.

9. Gustav Hellmann, *Die Anfänge . . .*, pp. 6-7.

10. *Ibid.*, p. 16.

11. For example, Descartes in 1647 proposed to take meteorological observations in concert with Mersenne. See René Descartes, *Oeuvres de Descartes*, ed. Chas. Adam et Paul Tonnery (Paris: 1903), 5:99.

12. A. Wolf, *A History of Science, Technology, and Philosophy in the 16th and 17th Centuries* (London: Allen & Unwin, 1935), p. 312.

13. Blaise Pascal, *The Physical Treatise of Pascal*, trans. I.H.B. and A.G.H. Spiers (New York: Columbia University Press, 1937), p. 116.

14. Wolf, *op. cit.*, p. 312.

15. Hellmann, in his article "The Dawn of Meteorology," states that he found 123 different series of meteorological observations belonging to the fifteenth, sixteenth, and seventeenth centuries. This number undoubtedly represents a small proportion of the total number of such observations throughout Europe.

16. For a detailed account of the meteorological activity of the Accademia del Cimento, see G. Hellmann, *Evangelista Torricelli, Esperienza dell'argento vivo. Accademia del Cimento, . . .* (Berlin: A. Asher & Co., 1897), pp. 11-22.

17. Thomas Sprat, *History of the Royal Society* (London: 1667), pp. 173-179.

18. Willis L. Webb, "Missile Range Meteorology," *Weatherwise* 16 (1963): 101.

19. Philippe De La Hire, "Observations of the quantity of water which fell at the observatory during the year 1709, with the state of the thermometer and barometer," *Memoirs of the Royal Academy of Science* (Paris: 1710) pp. 356-358.

20. *Ibid.*, pp. 356-357.

21. A. Wolf, *A History of Science, Technology, and Philosophy in the 18th Century* (New York: The Macmillan Co., 1939), p. 284.

22. James Jurin, "Invitatio ad Observationes Meteorologicas communi consilio instituendas," *Phil. Trans.* 32 (1723): 422-427.

23. H. Howard Frisinger, "Isaac Greenwood: Pioneer American Meteorologist," *Bulletin of the American Meteorological Society* 48, No. 4 (1967): 265-267.

24. Isaac Greenwood, "A New Method for Composing the Natural History of Meteors," *Phil. Trans.* 35 (1728): 390-402.

25. Greenwood was not the first to take regular meteorological observations on the American Continent. This honor apparently goes to Rev. John Campanius, who from 1644-1645 maintained a weather record at Swedes' Fort, near Wilmington, Del. See "A Chronological Outline of the History of Meteorology in the United States of North America," *Monthly Weather Review* (March 1909): 87.

26. Roger Pickering, "Scheme of a Diary of the Weather, together with draughts and descriptions of Machines subservient thereunto," *Phil. Trans.* 43 (1744): 6-7.

27. The Royal Society did not begin its own meteorological register until 1774, and then it only lasted seven years.

28. G.J. Symons, "The History of English Meteorological Societies, 1823 to 1880," *Quart. Jour. of the Roy. Meteor. Soc.* 7 (1881): 66-68.

29. In the back of his work *Physicae experimentales et geometricae . . .* (Lugduni Batavorum: 1729), Musschenbroek has included the printed record of the meteorological observations which he made at Utrecht in 1728, and in which he employed these symbols to represent meteorological phenomena.

30. Wolf, *18th Century*, p. 287.

31. J.H. Lambert, "Esposé de quelques Observations qu'on pourroit faire pour répandre du jour sur la Météorologie," *Nouveaux Memoires de l'Academie Royale des Science* (Berlin: 1771): 60-65.

32. Blaise Pascal, *The Physical Treatises of Pascal,* trans. I.H.B. and A.G.H. Spiers (New York: Columbia University Press, 1937), p. xvi.

33. *Ibid.,* pp. xvi-xvii.

34. Napier Shaw, *The Drama of Weather,* 2nd ed. (Cambridge: The University Press, 1939), p. 46.

35. R. Wootton, "The Physical Work of Descartes," *Science Progress* 21 (1927): 477. See also C. Adam, "Pascal et Descartes," *Revue Philosophique* 24 (1887), pp. 612-624; 25 (1888): 65-90; R. Duhem, "Le Pere Marin Mersenne et la pesanteur de l'air," *Revue Generale des Sciences* 17 (1906): 809-817.

36. Pascal, *op. cit.,* p. 164.

37. Here Torricelli is apparently referring to the experiments of Galileo.

38. Florian Cajori, "History of determinations of the heights of mountains," *Isis* 12 (1929): 499-500. For an English translation of the letters between Pascal and Perier concerning this experiment, see Forest R. Moulton and Justus J. Schifferes (eds.), *The Autobiography of Science* (New York: Doubleday and Company, Inc., 1953), pp. 148-152.

39. Florian Cajori, "History of determinations of the heights of mountains," *Isis* 12 (1929): p. 500.

40. Pascal, *op. cit.,* pp. 63-66.

41. Robert Boyle, *The Work of the Honourable Robert Boyle* (London: 1744), 1:97-104.

42. Another probable factor was that he was a close assistant in Boyle's experiments with the elasticity of air. See Margaret Espinasse, *Robert Hooke* (London: William Heinemann, Ltd., 1956), p. 46.

43. Robert Hooke, *Micrographia* (London: 1665), pp. 221-228.

44. *Ibid.,* p. 228.

45. Bernard De Lindenau, *Tables Barometriques* (Gotha: 1809), p. xxi.

46. Edme. Mariotte, *Oeuvres de M. Mariotte* (La Haye: 1740), 1:148-182.

47. Twelve successive applications of this process gave an altitude of nearly thirty-five miles. Mariotte stopped here as he had no evidence that air could be expanded beyond the degree of rarefaction which it would have at this altitude.

48. Florian Cajori. "History of determinations of the heights of mountains," *Isis* 12 (1929): 504.

49. Mariotte, *op. cit.,* pp. 174-175.

50. Edmund Halley. "On the height of the Mercury in the Barometer at different

Elevations above the Surface of the Earth; and on the Rising and Falling of the Mercury on the Change of Weather." *Philosophical Transactions of the Royal Society of London* (1686): 104-116.

51. Logarithms had been first invented in 1614 by John Napier. For a thorough account of the history of logarithms, see Cargill G. Knott, *Napier memorial volumes* (London: Longmans, Green and Co., 1915).

52. Halley, *op. cit.*, p. 109. The development by Halley of this formula can be summarized in modern mathematical notation as follows: by Boyle's Law pv = constant = 30 × 900 (Halley's constant). Thus, the cylinder of air reaching from sea-level to the place where the barometric reading is h, is

$$\int v\,dp = \int_h^{30} (30 \times 900)\frac{dp}{p} = \Big[(30 \times 900)\log p\Big]_h^{30}$$

$$= 30 \times 900 \times (\log 30 - \log h).$$

Changing from natural to common logarithms, by dividing by the modulus 0.434295, and by simplifying, Halley's formula is obtained.

53. Halley, *op. cit.*, p. 109.

54. Later in this work by Halley, he attempts to explain the reasons for the changes in the barometric readings at sea-level.

55. See H. Howard Frisinger, "Mathematicians in the History of Meteorology: The Pressure-Height Problem," *Historia Mathematica* 1 (1974): 263-286.

Chapter Eight

The End and the Beginning

The pattern of investigation evolved in the development of the pressure-height problem became the model for much of later meteorological development, including dynamic meteorology. In this pattern, the first step was the discovery of a basic atmospheric property. Next, attempts were made to establish a mathematical relationship between two or more of the involved variables. The last step was to express mathematically the corrections to the basic relationship, which had been empirically discovered. This method became the basis of dynamic meteorological development in the nineteenth and twentieth centuries.[1]

While most advances in dynamic meteorology occurred during the nineteenth and twentieth centuries, important first steps were made in the preceding century and a half. The most important questions were concerned with atmospheric motions, and the first to contribute to their answers was Edmund Halley (1656–1742).

In 1686, at the end of his work on the pressure-height problem, Halley remarked on the variations in barometric readings with different winds. The causes of these winds appeared to bother Halley, and it is not difficult to suppose that the problematic relation between barometric variations and wind led him to investigate the cause of winds. His results were published in the *Philosophical Transactions* in 1686.

Halley was the first to connect the general circulation of the atmosphere with the distribution of the sun's heat over the earth's surface. His theory was that heated air rises, and that the winds were caused by air flowing into these areas where the solar heat had caused the original air to rise. Halley used this theory to explain the easterly trade winds around the equator. The sun exposed the equatorial zone to a great amount of heat every day. The maximum heat follows the sun and, therefore, moves from east to west. The rarefaction caused by the rising air disturbs the equilibrium of the atmosphere, and a current of air will constantly follow the extreme of heat to restore the equilibrium

Edmund Halley
(1656–1742)

resulting in the trade winds. Halley included a world map (*Fig. 8.1*) which depicted the general wind circulation around the globe.[2] This, the earliest pictorial meteorological map.[3] shows a clear line of demarcation between the variable winds of the temperate zones on the one hand, and the more reliable winds of the tropics on the other, along a line which runs at around 30° both north and south of the equator.

Halley presented for the first time the concept of the winds as a general circulation of the air over the surface of the globe, generated by the distribution of the sun's heat over the earth's surface.[4] His idea of thermal convections has been the basis of many subsequent expositions on tropical atmospheric circulation and phenomena. By observing its motion, Halley was, therefore, a pioneer in the determination of the forces acting upon the atmosphere. For his significant contributions, he is often called the Father of Dynamic Meteorology.[5]

Nearly fifty years elapsed before the next step was taken by George Hadley, another Englishman (1685–1758). In 1735 the *Philosophical Transactions* published the result of Hadley's study of atmospheric motion,[6] in which he built upon Halley's theory, but for the first time made allowances for the earth's rotation.[7] Hadley first depicted the atmosphere surrounding the earth as revolving with it, so that there was no apparent motion of the atmosphere, and thus no wind. Then, he claimed, the action of the sun causes the air near the equator to be heated, rise and be replaced by cooler air from the north and south. The distance which these "drawn in" winds had to traverse in a revolution of the earth became greater as they moved toward the equator. Thus, the relative motion of these winds, to the revolving motion of the earth, resulted in the easterly winds near the equator (i.e., northeast winds on the north side of the equator, and southeast winds on the south side). The heated air about the equator rose and, being a fluid, spread itself north and south. Elevated high above the earth, this air eventually lost a great part of its heat, and thereby acquired sufficient density and weight to sink to the surface again. The relative velocity of the air, which originated about the equator, was greater than that of the earth, which caused a west wind in the middle latitudes. Hadley comments that all these winds must balance out, or there would be a change in the earth's motion about its axis.

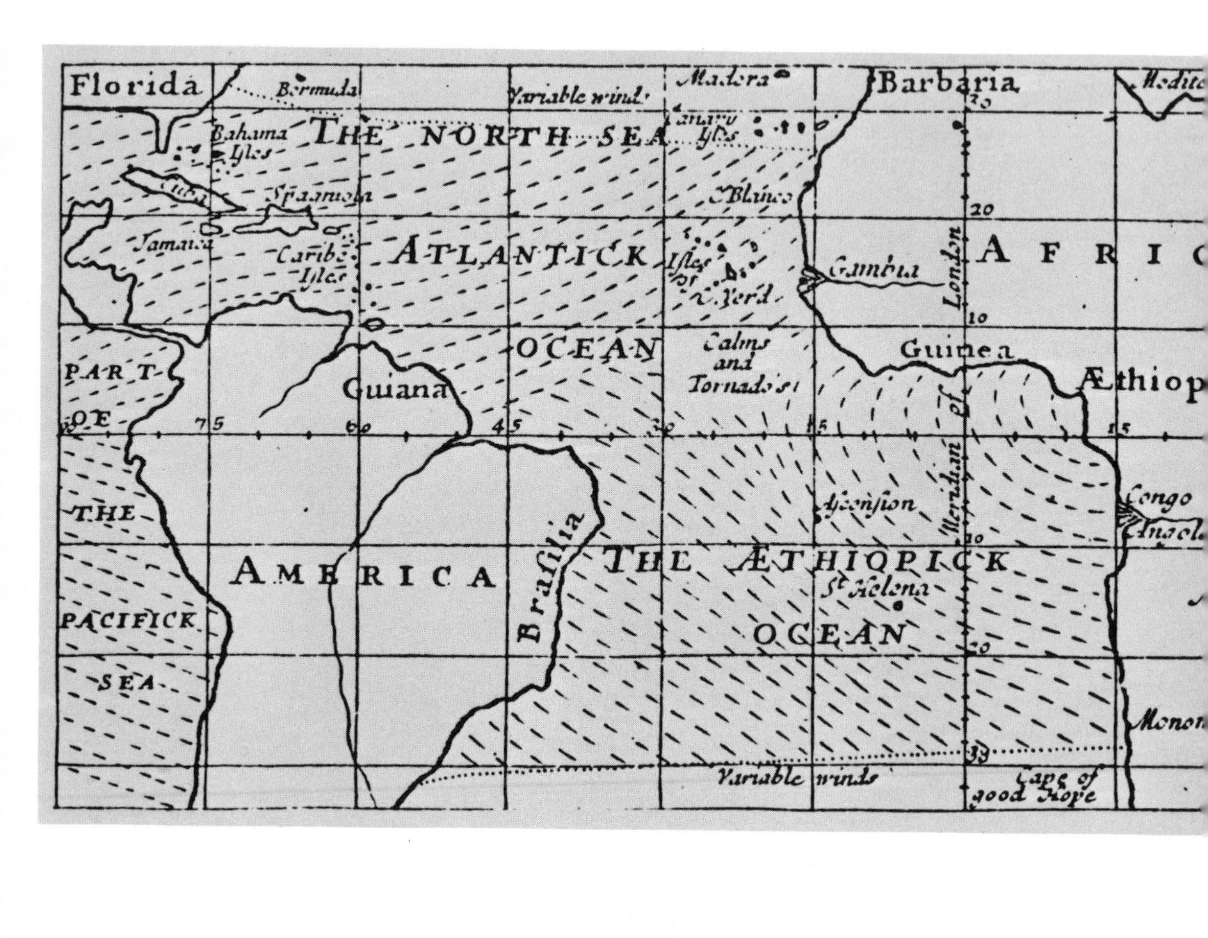

Florida
Bermuda
Variable winds
Madera
Barbaria
Bahama Isles
THE NORTH SEA
Canary Isles
Cuba
C. Blanco
Jamaica
Caribe Isles
ATLANTICK
OCEAN
Isles of C. Verd
Gambia
Meridian of London
Calms and Tornado's
Guinea
PART OF THE PACIFICK SEA
Guiana
AMERICA
Brasilia
THE ÆTHIOPICK OCEAN
Ascension
St Helena
Congo
Variable winds
Cape of good Hope

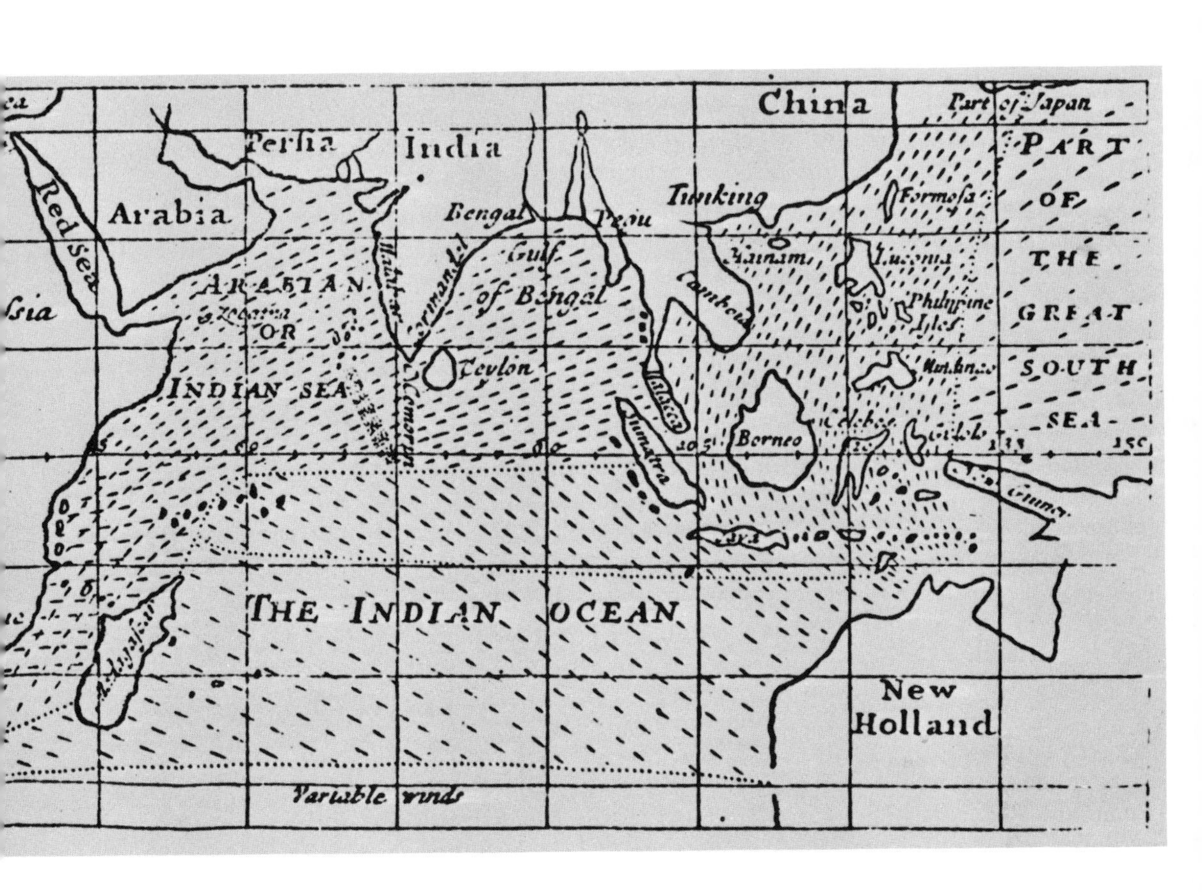

Halley's World Map of Wind Circulations
Fig. 8.1

Hadley's theory on the general cyclic circulation of air was generally accepted for nearly 200 years.[8] By the 1920's, however, most meteorologists realized that Hadley's theory was a simplification of a more complex natural phenomenon, as, indeed, recent rocket investigations have confirmed.[9] It is just this simplification, however, which made and still makes this theory so popular: It provided an easily understood, if not totally accurate, explanation for a vitally important atmospheric process.

Hadley's work is considered a milestone in the study of atmospheric motions because it was a starting point from which further study could proceed. In 1746 the Berlin Academy offered a prize for the best mathematical discussion of the atmosphere in which the earth was to be assumed to be covered by an ocean. Several noted European mathematicians and scientists responded, including Daniel Bernoulli. The Academy assembled an eminent group of scientists to judge the presented papers, headed by the noted mathematician, Leonard Euler.[10] They awarded the prize to the highly respected French mathematician and philosopher, Jean le Rond d'Alembert (1717–1783).[11]

In his prize-winning work, "Réflexion sur La Cause Generale Des Vents," dedicated to Frederick the Great, D'Alembert proposed a theory entirely different from Halley's and Hadley's.[12] D'Alembert asserted that the sun's heating of the earth's surface only affected the first few feet of the atmosphere, and could not be the primary cause of the winds. Rather, the winds were the result of the attracting forces of the sun and moon. With very intricate mathematical developments, employing Newton's recently discovered laws of gravitation, D'Alembert proceeded to form mathematical equations that described the movements of the atmosphere.

His treatise could be divided into three major parts. In the first, he assumed that the earth was a solid, smooth globe, covered by a thin layer of homogeneous, nonelastic air. Assuming also that the solar and lunar forces acting on the atmosphere were perpendicular to the earth's axis of rotation, D'Alembert developed equations expressing the resulting oscillations in the atmosphere, caused by the constantly changing net solar and lunar attraction forces.

In the second part, D'Alembert pretended that the earth was covered by an ocean, and that the atmosphere was not homogeneous, but elastic. After the velocity of the wind from the attraction

D'Alembert
(1717–1783)

force of one star (or planet) was found, the velocity arising from the attraction forces of two stars were determined. The third part of D'Alembert's treatise was devoted to the effect such impediments as mountains had upon the winds.

D'Alembert harbored no delusions that his work provided a final solution to the question of the cause and calculations of atmospheric winds, and realized that there were a great many factors, other than the solar and lunar attraction forces, which effected the winds. He correctly claimed, however, that the lack of knowledge about the dynamic properties of the atmosphere made the inclusion of such factors in any mathematical calculation impossible. Thus, D'Alembert believed that, under the guidelines established by the Berlin Academy, his theory and calculations were the best available. Like Halley, D'Alembert included in his treatise a map of the winds in the lower latitudes (*Fig. 8.2*).

D'Alembert's contribution to the development of dynamic meteorology has generally been overlooked by historians probably because his "solar and lunar force" theory has been proved invalid. He had tried to establish a reasonable analogy between the attraction forces of the sun and moon on the oceans causing tides, and these forces on the atmospheric "fluid" causing winds. Unfortunately, this analogy did not hold true.

There are, nevertheless, several reasons why his work was important for the later development of dynamic meteorology. In the first place, the mere attempt by a noted mathematician and scientist to treat a section of meteorology mathematically was instrumental in the slow but steady acceptance of meteorology as a legitimate science. D'Alembert was the first to attempt to express mathematically the motions of the atmosphere. His work was one of the first manifestations of a trend in meteorology which became more and more prevalent in the nineteenth century.[13]

Halley, Hadley, and D'Alembert provided the initial inducements to the study of dynamic meteorology by nineteenth-century scientists. The basic physical and mathematical tools needed in the study of the atmosphere were also extensively developed by 1800. It is this development that we now wish to consider.

The equations governing the atmosphere are (1) the equations of motion, (2) the equation of continuity, (3) the first law of thermodynamics, (4) the equation of state or the ideal gas law, and (5) the hydrostatic equation.[14] In many applications in

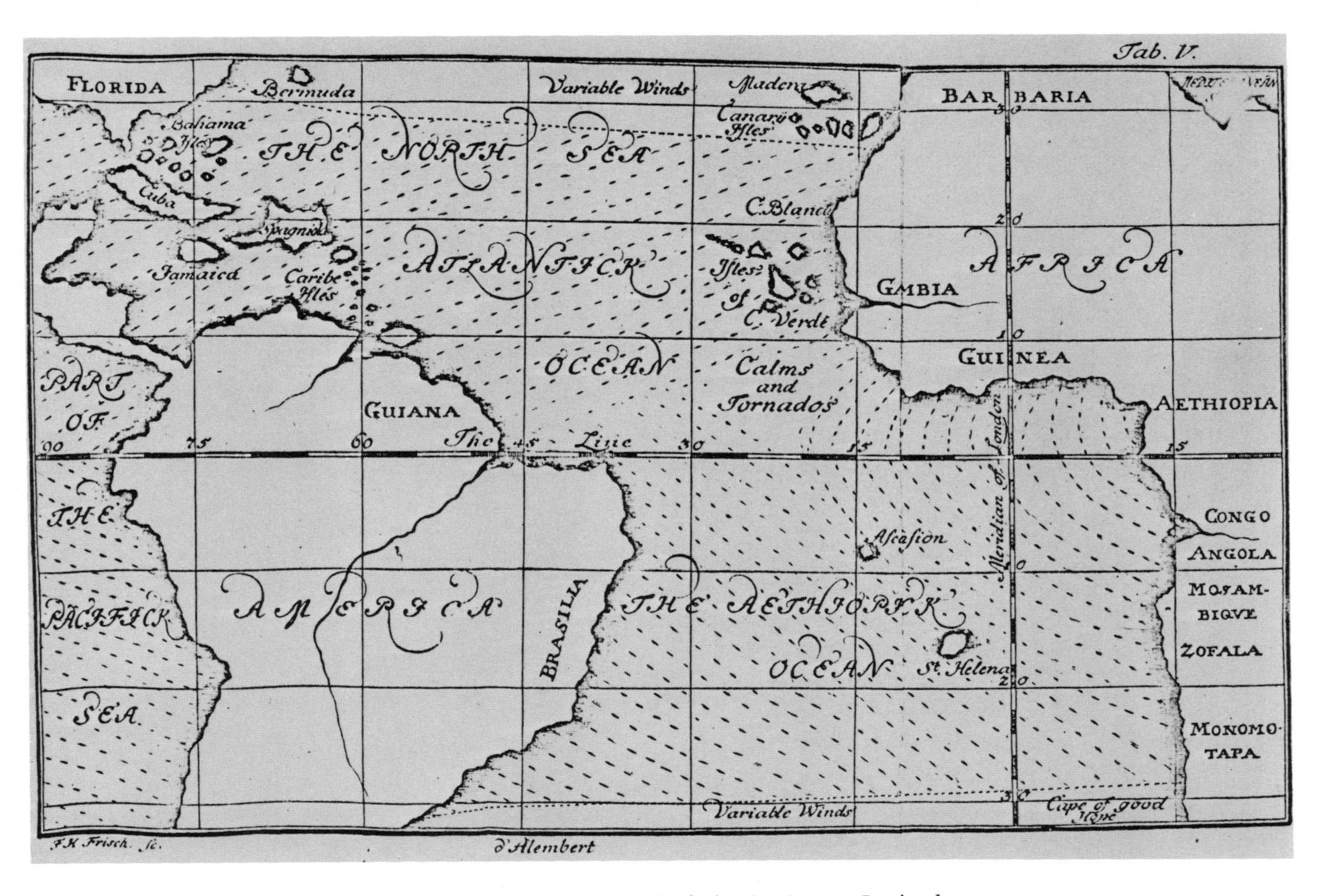

D'Alembert's Map of the Winds in the Lower Latitudes

Fig. 8.2

meteorology, the three equations of motion (1) are reduced to two equations for horizontal wind components and the hydrostatic equation for the vertical equation. The Coriolis force, which was not formulated until the nineteenth century,[15] appears in the equations of motion (1).

The first law of thermodynamics, which is the law of the conservation of energy, was formulated in the nineteenth century and is beyond the scope of this book.[16] The equation form of the ideal gas law is

$$(1) \quad p\alpha = RT$$

where α is the specific volume under pressure p and temperature T; R is the specific gas constant for the gas being considered. This law is a combination of Boyle's and Charles' laws.

As was noted in Chapter VI, Boyle's law states that under constant temperature the volume of a gas is diminished in the same ratio as the pressure upon it is increased, or in equation form

$$(2) \quad p\alpha = C$$

where the constant C depends on the fixed temperature of the gas. Around 1787, Jacques Alexandre Cesar Charles (1746-1823) discovered what is known as Charles' Law,[17] which states that at fixed pressure the volume of a gas is directly proportional to the absolute temperature. Charles did not publish his discovery, and it was not until 1802 that Joseph Gay-Lussac (1778-1850) published papers that included this law.[18]

The behavior of real gases follows Boyle's and Charles' laws only approximately. During the early part of the nineteenth century, however, the concept of an "ideal gas" in which the above two laws followed exactly was introduced. This concept allowed Boyle's and Charles' laws to be combined to obtain the ideal gas law as given in equation (1). In 1813 Amerigo Avogadro (1776-1856) used his law—at equal pressures and temperatures a molecular weight of two different gases will occupy the same volume—to obtain another form of the ideal gas law:

$$(3) \quad p = \frac{R^*}{m} T$$

where m is the molecular weight of the gas, and R^* is the universal gas constant.[19] Thus, although the ideal gas law was not formulated until early in the nineteenth century, its foundations were laid prior to 1800.

The origin of the hydrostatic equation is not so clearly defined. The study of hydrostatics has slowly progressed since ancient Greece, when Archimedes founded the science.[20] A needed ingredient for the formulation of the hydrostatic equation is Pascal's Law, which states that two points at the same elevation in the same continuous mass of fluid at rest have the same pressure.[21] Although Pascal's work, a systematization of hydrostatics and aerostatics, was an achievement of major importance,[22] the mathematical concepts of partial and total derivatives needed for the hydrostatic equation were a product of the eighteenth century.

Let us briefly look at this equation. The partial derivative expressions

$$\frac{\partial p}{\partial x} = 0, \quad \frac{\partial p}{\partial y} = 0$$

are one of the several forms of Pascal's Law. For the vertical direction, however, the body force, or weight, of the element must be included. This results in the partial derivative

$$(4) \quad \frac{\partial p}{\partial z} = -g\rho.$$

Inasmuch as the pressure is a function of the vertical direction only when Pascal's law is assumed valid, the partial derivative (4) becomes a total derivative of pressure with respect to vertical distance. Then the hydrostatic equation reads

$$(5) \quad \frac{dp}{dz} = -g\rho, \quad \text{or} \quad dp = -g\rho dz.$$

The hydrostatic equation (5) first appeared in the tenth book of Pierre Simon Laplace's monumental work *Mécanique Céleste*, initially in a discussion of the refraction of light,[23] and later in his development of a pressure-height equation.[24] Laplace introduced the hydrostatic equation (5) by stating: "The decrement of the pressure p, produced by the elevation dz, is evidently equal to the small column of air ρdz, multiplied by the negative of the gravity."[25] This use of 'evidently' and his earlier statements indicates that the equation was *primarily empirically* derived.

We do not know if Laplace was the first person to formulate the hydrostatic equation. The problem stems from Laplace's refusal to avow his indebtedness to others. According to E.T. Bell, "Laplace stole outrageously, right and left, whenever he could lay his hands on anything of his contemporaries and predecessors which he could use."[26] Regardless of who first formulated the hydrostatic equation, however, the principle behind it was known by 1800.

In fairness to Laplace (1749–1827), it should be noted that he was certainly capable of its discovery. He was one of the outstanding scientists of the 1750–1850 period, often called "the Newton of France."[27] Besides his contribution to hydrostatics and the pressure-height problem, Laplace made several other indirect contributions to the development of meteorology: the Laplacian operator ∇^2 (a partial differential operator of calculus that expresses how a mathematical or physical variable, such as pressure, changes with respect to spatial coordinates), his removal of the discrepancy between the actual and the Newtonian velocity of sound, his experiments with Lavoisier on specific heats, his invention of the ice-calorimeter, and his work on the development of electricity by evaporation. In addition, Laplace worked with other European scientists in establishing a network of weather observing stations from 1800–1815.[28]

We now come to the basic equations governing atmospheric motion; the equations of motion, and the equation of continuity. As the mathematical formulation of the equation of continuity was an apparent by-product of the study of the equations of motion, the two developments are considered together. The history of the equations of motion can be subdivided into two periods; the early development of the theories of mechanics culminating with Newton's Second Law, and the mathematical development of the theory of partial differential equations with their application to fluid dynamics in the eighteenth century.

The foundations of mechanics were laid in the 3rd century B.C. by the Greek mathematician, Archimedes, who derived formulas for the equilibrium of simple levers and centers of gravity.[29] The derivations were restricted to parallel forces. The treatment of nonparallel forces presented difficulties that were not overcome until force was conceived as a vector quantity with both magnitude and direction, describable by a directed line segment.

This did not occur until over 2,000 years later when the sixteenth-century Dutch mathematician Simon Stevin (1548–1620) solved the lever problem with nonparallel forces.[30] By observation and intuition alone, Stevin demonstrated the equilibrium of bodies on a double inclined plane. He showed how to add force vectors by constructing a parallelogram using the force vectors as the sides.[31] Stevin's work gave a new impetus to the study of mechanics and led to the important work of Galileo.

Many early Greek scientists, particularly Aristotle, tried to explain the behavior of moving bodies. In these attempts they were unsuccessful: first, because they had no satisfactory means for measuring distance or time and, consequently, were unable to check their formulas experimentally; and second because they laboured under the false assumption that force was necessary to maintain motion, rather than to change its direction or magnitude. It was left to Galileo to resolve these problems and thereby found dynamics.

One of Galileo's important contributions to science dealt with the motion of falling bodies.[32] He disproved the Aristotelian theory that heavy bodies fall faster than light bodies, and proposed a new theory that the height fallen would be proportional to the square of the time and independent of the weight.[33] To verify it, Galileo experimented with a smooth ball rolling down an inclined plane; he measured the time of the motion by weighing the water that flowed through a small hole at the bottom of a large water-filled tank.

Galileo developed the equation of motion of a projectile and showed that, neglecting air resistance, the curved path was the result of two independent motions—a horizontal at constant speed, and a vertical at a speed varying with the force of gravity. In this case he was utilizing the principle of compounding motions in accordance with the parallelogram law for force vectors introduced by Stevin.

Another theory of Galileo, that a body in motion and free from external forces would keep on moving at a constant speed in a straight line, anticipated Isaac Newton's first law of motion, the law of inertia.[34] Galileo was the first to recognize that it was acceleration, and neither the velocity nor the position, that was determined by the external forces. His explanation of the path of projectiles indicates that Galileo had also grasped Newton's second law of motion.[35] Galileo, however, did not fully realize the third law of motion, the law of the equality of action and reaction, though he corrected some errors of Aristotle.

Further advances in the science of mechanics were made in the seventeenth century by John Wallis, Christopher Wren, and particularly Christian Huygens.[36] Huygens developed the equations of motion of the pendulum and invented the pendulum clock. He was the first to obtain the acceleration of gravity by pendulum observations, and also the first to formulate the term now known as the moment of inertia (the measure of the inertia of a body in rotary motion).

The works of Galileo and Huygens were instrumental in freeing mechanics from the scholastic discipline. Basic problems like the motion of projectiles in the vacuum and the oscillations of a compound pendulum had been solved. Nevertheless, the task of putting these principles together into an organized whole remained. This was the contribution of Newton, who set his seal on the foundation of classical mechanics at the same time that he extended its field of application to celestial phenomena. Newton's work in mechanics appeared in his famous treatise *Philosophal Naturalis Principia Mathematica* (1687).[37]

The *Principia* is divided into three books. In a prefatory section Newton first defines concepts of mechanics such as force, inertia, and momentum. Next, he states his three famous laws of motion:

> Law I (Law of Inertia). Every body continues in its state of rest, or of uniform motion in a straight line, unless it is compelled to change that state by forces impressed upon it.
>
> Law II. The change of velocity is proportional to the force impressed, and is made in the direction of the straight line in which that force is impressed, and is inversely proportional to the mass of the body. (i.e. $F = Ma$).
>
> Law III. To every action there is always opposed an equal reaction.

These three laws form the basis of modern mechanics. In particular, the second law is very important for from it all of the basic equations of dynamics can be derived by procedures developed in calculus. These procedures were mostly developed in the eighteenth century, and it is this second period in the development of the equations of motion which we now wish to consider.

Leonhard Euler
(1707-1783)

In addition to the important contributions to mechanics, Newton's *Principia* also contained the first formulation of calculus.[38] This very powerful mathematical tool stimulated extensive efforts by eighteenth-century mathematicians to develop the concepts of calculus, and to apply the calculus to the solution of physical problems. One major outgrowth of these efforts was the creation of the concept of partial differential equations.[39] Systems of partial differential equations arose first in the eighteenth-century work on fluid dynamics and hydrodynamics. The work on compressible fluids, air in particular, sought to analyze the propagation of sound, the action of air on sails of ships, and the design of windmill vanes. Several mathematicians contributed to both the development of the theory of partial differential equations and to its applications. The one man, however, who stood above everyone else, and for whom the science of meteorology is most in debt, was Leonhard Euler (1707–1783).

Euler took a sincere interest in the new science of meteorology, particularly the pressure-height problem which he studied under various hypotheses about temperature and gravity.[40] Despite blindness and old age, he was one of the scientists to answer the call of the meteorological society of the Palatinate to take daily weather observations. In his famous *Lettres à une Princesse d'Allemagne,* Euler discussed and summarized the meteorological knowledge.[41] His interest and full support were undoubtedly important stimulants to the scientific study of meteorology.

Euler's major contribution to the emerging science was to use Newton's second law and the concepts of partial differential equations to develop in 1755 the equations of fluid flow for perfect (nonviscous) compressible and incompressible fluids.[42] Euler regarded fluid as a continuum and the particles as mathematical points. The force acting on a small volume of the fluid was considered subject to the pressure p, density ρ, and external forces with components R, S, T per unit mass. The components u, v, and w of the fluid velocity were expressed at every point in the fluid by $u = u(x,y,z,t)$, $v = v(x,y,z,t)$, and $w = (x,y,z,t)$. Now

$$(6) \quad du = \frac{\partial u}{\partial x}\,dx + \frac{\partial u}{\partial y}\,dy + \frac{\partial u}{\partial z}\,dz + \frac{\partial u}{\partial t}\,dt.$$

In the period of time dt, the particle at (x,y,z) traveled a distance udt in the x-direction, vdt in the y-direction, and wdt in the z-direction. Thus, the actual changes dx, dy, and dz in equation (6) are given by these quantities in the following equations:

$$(7)\quad \begin{aligned} \frac{du}{dt} &= u\frac{\partial u}{\partial x} + v\frac{\partial u}{\partial y} + w\frac{\partial u}{\partial z} + \frac{\partial u}{\partial t} \\ \frac{dv}{dt} &= u\frac{\partial v}{\partial x} + v\frac{\partial v}{\partial y} + w\frac{\partial v}{\partial z} + \frac{\partial v}{\partial t} \\ \frac{dw}{dt} &= u\frac{\partial w}{\partial x} + v\frac{\partial w}{\partial y} + w\frac{\partial w}{\partial z} + \frac{\partial w}{\partial t} \end{aligned}$$

By calculating the forces acting on the particle at the point (x,y,z) and applying Newton's second law, Euler obtained his following famous equations of motion:

$$(8)\quad \begin{aligned} \frac{du}{dt} &= R - \frac{1}{\rho}\frac{\partial p}{\partial x} = \frac{\partial u}{\partial t} + \bar{v}\cdot\nabla u \\ \frac{dv}{dt} &= S - \frac{1}{\rho}\frac{\partial p}{\partial y} = \frac{\partial v}{\partial t} + \bar{v}\cdot\nabla v \\ \frac{dw}{dt} &= T - .\frac{1}{\rho}\frac{\partial p}{\partial z} = \frac{\partial w}{\partial t} + \bar{v}\cdot\nabla w \end{aligned}$$

Euler also generalized the differential equation of continuity

$$\frac{\partial u}{\partial x} + \frac{\partial v}{\partial y} + \frac{\partial w}{\partial z} = 0$$

(this says that no matter is destroyed or created during motion), and obtained its modern form

$$(9)\quad \frac{\partial \rho}{\partial t} + u\frac{\partial \rho}{\partial x} + v\frac{\partial \rho}{\partial y} + w\frac{\partial \rho}{\partial z} + \rho\left(\frac{\partial u}{\partial x} + \frac{\partial v}{\partial y} + \frac{\partial w}{\partial z}\right) = 0$$

This concept had been earlier expressed in words by Francis Bacon, Robert Boyle, Newton and Hadley;[43] and first expressed as a differential equation by D'Alembert.[44]

Euler's development of the equations of motion was a great step in modern fluid dynamics.[45] In this development he made the transition from parcels to points. No longer was it necessary to follow each parcel to calculate the forces upon it and thus deduce its accelerations. Instead the calculation could be performed in a fixed coordinate system where rates of change at a point could be determined.

This transition from parcels to points is a crucial relation in numerical weather prediction. By replacing the necessity of following parcels on their trajectories with non-linearity ($\bar{v} \cdot \boldsymbol{\nabla}$), the history of fluid dynamics has become the study of the non-linear equations of motion.

Thus, the year 1800 marked both an end and a beginning in the history of meteorology. The 2000 year period in which meteorology struggled from speculation into an experimental science reached its climax in the eighteenth century with the development of the basic meteorological instruments and the basic set of dynamic equations. By 1800 all but one of the fundamental tools for the study of the atmosphere had been developed, in particular Euler's equations of motion. The post-1800 development has been basically the applications of these equations.

The one missing ingredient to the complete set of equations governing the atmosphere was the first law of thermodynamics which was not formulated until well into the nineteenth century. This law is quite crucial for meteorology. The constant density set of equations or the horizontal, hydrostatic set can be used to model some aspects of atmospheric flow. The motions of the atmosphere, however, are a response to thermal forcing; they are a thermodynamic phenomenon. Thus, although scientists were ready by the beginning of the nineteenth century to study the mechanical aspects of atmospheric flow, they were not quite ready to study its thermodynamic properties. The study of these thermodynamic properties, along with the study of the atmospheric applications of the mechanical equations of motion, was the main task of nineteenth-century meteorology.

By the close of the eighteenth century the main arch of dynamic meteorology had largely been built; it only remained to slip the keystone (first law of thermodynamics) into place. Thus, it can be truly said that the eighteenth century marked the beginning of the development of the science of the atmosphere into a separate and dynamic branch of the physical sciences.

References

1. A classic in the attempts to express the motions of the earth's atmosphere in the form of mathematical equations was L.F. Richardson, *Weather Prediction by Numerical Process* (Cambridge: Cambridge University Press, 1922).

2. Edmund Halley, "An Historical Account of the Trade Winds and Monsoons, Observable in the Seas between and near the Tropics; with an Attempt to Assign their Physical Cause," *Phil. Trans.* 16 (1686): 153–168.

3. Napier Shaw, *Manual of Meteorology* (Cambridge: Cambridge University Press, 1926), 1:258.

4. Charles Singer, *A Short History of Scientific Ideas to 1900* (Oxford: The Clarendon Press, 1959), p. 321.

5. Shaw, *Manual*, p. 122.

6. George Hadley, "Concerning the Cause of the General Trade Winds," *Phil. Trans.* 34 (1735): 58–62.

7. Julius Von Hann, *Lehrbuch Der Meteorologie*, 4th ed., (Lepizig: Herm. Tauchnitz, 1926), p. 437.

8. For example, see William L. Donn, *Meteorology with Marine Applications* (New York: McGraw-Hill Book Co., Inc., 1946), p. 149.

9. See W.W. Kellogg, "Report on Symposium on Meteorological Rockets," *Proceedings of the First International Symposium on Rocket and Satellite Meteorology* (New York: John Wiley & Sons, Inc., 1963), pp. 9–14.

10. Paul H. Fuss, *Correspondance Mathematique et Physical De Quelques Célèbres Géomètres Du XVII eme Siecle* (St. Petersburg: 1843), 1:368–372.

11. Cleveland Abbe, trans., *The Mechanics of the Earth's Atmosphere*, 3rd collection (Washington: Smithsonian Institution, 1910), p. 1.

12. Jean Le D'Alembert, *Reflexions sur La Cause Generale Des Vents* (Berlin: 1747).

13. Cleveland Abbe, "The Progress of Science as Illustrated by the Development of Meteorology," *Smithsonian Institution Annual Report* (1907): 299.

14. See Seymore L. Hess, *Introduction to Theoretical Meteorology* (New York: Henry Holt and Company, 1959), p. 215.

15. Coriolis force is named after a French mathematican G.G. deCoriolis (1792–1843). He published his work on the effect of the earth's rotation in 1835. There is some question as to the appropriateness of assigning Coriolis' name to this deflecting force. See H.E. Landsberg, "Why indeed Coriolis?," and Harold L. Burstyn, "The deflecting force and Coriolis," *Bulletin of the American Meteorological Society* 47, No. 11 (November 1966): 887–891.

16. For a discussion of the historical development of this law in the nineteenth century, see Florian Cajori, *A History of Physics* (New York: Dover Publications, Inc., 1962), pp. 218–223.

17. *Ibid*, p. 206.

18. *Ibid.*

19. Sir William Dampier. *A History of Science* (New York: The Macmillan Company, 1935), p. 229.

20. Morris Kline, *Mathematical Thought from Ancient to Modern Times* (New York: Oxford University Press, 1972), pp. 165-166.

21. Blaise Pascal, *The Physical Treatise of Pascal,* trans. I.H.B. and A.G.H. Spiers (New York: Columbia University Press, 1937), pp. 156-292.

22. W.E. Knowles Middleton, *The History of the Barometer* (Baltimore: The Johns Hopkins Press, 1964), p. 53.

23. Pierre Simon Laplace, *Mécanique Célest,* trans. Nathaniel Bowditch (Boston: Hillard, Gray, Little and Wilkins, 1829-39), 4:478. The original work was published in parts from 1798 to 1825.

24. *Ibid.,* p. 565. For a discussion of Laplace's pressure-height equation see H. Howard Frisinger, "Mathematicians in the History of Meteorology: The Pressure-Height Problem," *Historia Mathematica.*

25. *Ibid.,* p. 478.

26. E.T. Bell, *Men of Mathematics* (New York: Simon and Schuster, 1962), p. 174.

27. Sir William Napier Shaw, *Manual of Meteorology* (Cambridge: The University Press, 1926), 1:130.

28. Harvey A. Zinszer, "Meteorological Mileposts," *Scientific Monthly,* 58 (1944): 262.

29. For a complete discussion on the works and methods of Archimedes, see T.L. Heath(ed.), *The Works of Archimedes* (New York: Dover Publications, Inc.)

30. Charles Singer, *A Short History of Scientific Ideas to 1900* (London: Oxford University Press, 1959), pp. 223-224.

31. Rene Dugas, *A History of Mechanics,* trans. J.R. Maddox (Switzerland: Editions Du Griffon, 1955), pp. 124-127.

32. Galilei Galileo, *Dialogues Concerning Two New Sciences,* trans. Henry Crew and Alfonso DeSalvio (Chicago: Northwestern University Press, 1939), pp. 64-68.

33. *Ibid.,* p. 265.

34. *Ibid.,* p. 244.

35. *Ibid.,* p. 250.

36. See Rene Duga, *op. cit.,* pp. 172-199.

37. Isaac Newton, *Philosophial Naturalis Principia Mathematica,* 3rd ed., trans. Andrew Motte, ed. Florian Cajori (Berkeley: University of California Press, 1946).

38. See Carl B. Boyer, *The History of the Calculus and Its Conceptual Development* (New York: Dover Publications, Inc. 1959), pp. 187–223.

39. For a thorough discussion of the development of partial differential equations in the eighteenth century see Morris Kline, *op. cit.*, pp. 502–543.

40. Leonhard Euler, *Die Gesetze des Gleichgewichts und der Bewegung Flussiger Korper*, trans. H.W. Brandes (Leipzig: 1806), pp. 53–86.

41. Leonhard Euler, *Letters of Euler on Different Subjects in Natural Philosophy Addressed to a German Princess*, trans. David Brewster, 2 vols. (New York: Harper & Brothers, 1872). Throughout this collection of letters by Euler are comments on several meteorological phenomena, and properties of the atmosphere. In letters X, XI, and XII, written in May 1760, Euler summarizes the then known knowledge of the rarefaction and elasticity of the air, of gravity, and of the barometer, in the letters XXXVI–XXXIX, written in August 1761, the author presents the theories on thunder and lightning, held by the ancient philosophers, and Descartes's theory. Due to their lack of knowledge of electricity, these previous theories were "a mass of absurdity." According to Euler, the stormy clouds are extremely electrical, and thunder and lightning are the result of a huge electrical spark set off by these clouds.

42. Leonhard Euler, "General Principles of the Motion of Fluids," *Hist. de l'Acad. de Berlin* 11 (1755): 274–315.

43. Singer, *op. cit.*, pp. 332–333.

44. Kline, *op. cit.*, p. 541.

45. It should be noted that Euler's equations are not the final ones for hydrodynamics. He neglected viscosity which was not introduced until 70 years later by Navier and Stokes. See Kline, *op. cit.*, pp. 696–698.

Index

PICTURE CREDITS

Fig. 1.2: From Howard Eves, *An Introduction to the History of Mathematics,* Third Edition. Copyright © 1953 by Howard Eves. Copyright © 1964, 1969 by Holt, Rinehart and Winston, Inc. Reprinted by permission of Holt, Rinehart & Winston

Fig. 3.1: From J.H. Breasted, *Ancient Times,* Ginn & Co., 1916

Fig. 3.2: The National Museum, Athens

Fig. 4.5: From Henry C. Bolton, *Evolution of the Thermometer 1592–1743,* The Chemical Publishing Co., 1900

Fig. 4.6: From *Philosophical Transactions of the Royal Society of London,* 1701

Figs. 5.3, 6.1: From Robert Hooke, *Micrographia,* 1665

Figs. 5.4, 5.5, 5.6, 6.6: From W.E. Knowles Middleton, *Invention of the Meteorological Instruments,* The Johns Hopkins Press, 1969. Reprinted by permission of The Johns Hopkins Press

Fig. 6.4: From A. Wolf, *A History of Science, Technology and Philosophy in the 18th Century,* The Macmillan Company, 1938. Reprinted by permission of Macmillan Publishing Co., Inc.

Fig. 6.5: From the *Quarterly Journal of the Royal Meteorological Society,* 1882

Fig. 6.8: From W. Derham, *Philosophical Experiments and Observations of Hooke,* 1726

Fig. 7.1: From Thomas Sprat, *History of the Royal Society,* 1702

Fig. 7.2: Original from *Philosophical Transactions,* 1729; reprint from the collection of the author

Fig. 7.3: From *Philosophical Transactions,* 1744

Fig. 7.4: From *Nouveauz Memoires de L'Academie Royale Des Sciences,* 1771

Fig. 8.1: From *Philosophical Transactions,* 1686

Fig. 8.2: From D'Alembert, *Reflexions Sur La Cause Generales Des Vents,* 1747

All portraits courtesy of History of Science Collections, University of Oklahoma Libraries, except for Thales of Miletus (p. 13), coirtesy of Ny Carlsberg Glyptotek, Copenhagen.

All other illustrations courtesy of the author's private collection.

Library of Congress Cataloging in Publication Data

Frisinger, H Howard, 1933-
The early history of meteorology.

Bibliography: p.
1. Meteorology--History. I. Title.
QC855.F74 551.5'09 75-31681
ISBN 0-88202-036-6